ENCE BURNS served as corporate vice president of research, opment and planning at General Motors, where he oversaw s advanced technology and innovation programmes as well as orate strategy. He was also a professor of engineering practice the University of Michigan and led the Program for Sustainable bility at Columbia University. He has served as an adviser to e Google self-driving car project (now Waymo) since 2011 and a member of the US National Academy of Engineering. He lives Franklin, Michigan.

Praise for *Autonomy*

you want a glimpse of how the future is being engineered today, is no better book'

JEFFREY SACHS, author of *The End of Poverty*

ntertaining and accessible account of the biggest disruption history of the auto industry, and indeed the entire transpor- industry'

——ICK WAGONER, former chairman and chief executive officer, General Motors

Plea on th

ider's view into the thrilling who-will-win-it race to invent

To re trol driverless cars — and the radically altered future that

or co low in their wake'

Your   ROBIN CHASE, cofounder, Zipcar, and author of *Peers Inc*

nd entertaining insider's account . . . required reading for nterested in the future of mobility'

ROGER MARTIN, co-author of *Playing to Win*

'Tells the remarkable story of innovators who are changing trans- portatio

CL.   or's Dilemma*

'Takes us inside the auto industry as it is today – and what may be a very different industry tomorrow'

DANIEL YERGIN, author of *The Prize* and *The Quest*

'Essential reading'
The Times

Autonomy

The Quest to
Build the Driverless
Car – and How
It Will Reshape
Our World

LAWRENCE BURNS

WITH CHRISTOPHER SHULGAN

WILLIAM
COLLINS

William Collins
An imprint of HarperCollins*Publishers*
1 London Bridge Street
London SE1 9GF

www.WilliamCollinsBooks.com

First published in Great Britain by William Collins in 2018
First published in the United States in 2018 by Ecco

This William Collins paperback edition published in 2019

1

A catalogue record for this book is
available from the British Library

ISBN 978-0-00-830210-8

Designed by Michelle Crowe
Printed and bound by CPI Group (UK) Ltd, Croydon, CR0 4YY

MIX
Paper from
responsible sources
FSC® C007454

This book is produced from independently certified FSC™ paper
to ensure responsible forest management.

For more information visit: www.harpercollins.co.uk/green

To engineers, who make what's possible real

One new idea leads to another, that to a third and so on through a course of time, until someone, with whom no one of these ideas was original, combines all together, and produces what is justly called a new invention.

—Thomas Jefferson

Contents

Autonomy

Introduction

THE PROBLEM WITH CARS

I can't understand why people are frightened of new ideas.
I'm frightened of the old ones.

—JOHN CAGE

The way we get around is changing. For the first time in 130 years, we're in the midst of a major transformation in automobile transportation. In contrast to the personally owned, gasoline-powered, human-driven vehicles that have dominated the last century, we're transitioning to mobility services based on electric-powered and driverless vehicles, paid for by trip or through subscriptions.

What does this mean? Soon, many of us will no longer need to own or drive a car. Instead, we will rely on services that safely and conveniently use autonomous vehicles to take us where we want to go. The providers will manage every aspect of our transportation experiences, from vehicle parking to cleaning and maintenance to recharging. The hassles of car ownership will be eliminated. No longer will we need to shop for, finance and insure a car, or spend our time driving, parking or pumping gas. Traffic will be less of a headache. And we will be able to choose between riding in shared vehicles that also serve others or paying more to have an exclusive autonomous "valet" that not only takes us door to door when we want, but also can be dispatched to run errands or transport family and friends.

Summoning a ride will happen with the touch of an app. The vehicle that arrives won't have a steering wheel or gas and brake pedals. Most trips will happen in electric vehicles tailored to comfortably seat two people, since most trips we make happen solo or with just one other person. All this—and transportation is going to cost us just a fraction of what it ever did before.

This book chronicles the origins of the coming transformation. The changes I describe use current technology to solve the transportation problem in a different way. We don't often consider transportation a problem, but it is. Without giving it much thought, every day, every one of us considers the dilemma of how to get where we want to go when we want to be there. We come up with various solutions. For more than a century, the predominant solution in North America has been the personally owned, gas-powered, human-operated automobile. But that particular answer has caused numerous issues.

Today in the United States, 212 million licensed drivers own 252 million light-duty vehicles and drive 3.2 trillion miles a year, burning more than 180 billion gallons of fuel as they do. The emissions of cars and trucks amount to a fifth of the greenhouse gases created in the United States. And the distance we travel by automobile is growing, with the number of vehicle miles traveled increasing about 50 percent from 1990 to 2016.

We've structured our transportation in such a manner that most working adults believe that owning and maintaining their own vehicles is integral to their full participation in contemporary society. Yet, American automobiles sit unused about 95 percent of the time.

When we *do* drive those vehicles, they're terribly inefficient. More than 95 percent of the automobiles sold in the United States today are propelled by internal combustion engines that use gasoline. Less than 30 percent of the energy from the gasoline you put in your car is used to move it down the road. The rest of the energy is wasted as heat and sound, or used to power accessories like headlights, radios and air conditioners. Because typical vehicles weigh around 3,000 pounds and typical people weigh around 150 pounds, only about 5 percent of the gasoline energy translated into motion is used to move the driver, which amounts to just 1.5 percent of the total energy in gasoline.

Such inefficiencies arise because we purchase automobiles that are massively overbuilt for the purposes we most frequently use them. Waymo CEO John Krafcik calls this the "occasional-use imperative." Think about it. In the United States, 85 percent of personal travel is by automobile. Average occupancy of 1.7 people per mile falls to just 1.1 in vehicles conducting work commutes. Average speed in congested cities can run as low as 12 mph. And yet the cars, trucks and SUVs we drive have enough room for at least five adults, with engines so powerful many can travel at 120 mph and beyond. "The mix of cars on our nation's roadways is completely messed up," Krafcik observes.

These overbuilt vehicles are dangerous, because they're heavy. The World Health Organization estimates that auto crashes around the world kill 1.3 million people a year. In 2016 alone, 37,461 Americans were killed in auto crashes, contributing to make unintentional injuries the leading cause of death for Americans in the first half of life.

Using your vehicles just 5 percent of the time means that you have to figure out a place to store them the other 95 percent. So, you need to devote a good chunk of your home to a garage (and driveway), and not only that—where you work has to reserve space for your car, too. As does your favorite shopping mall, your dentist's office, the stadium for your favorite sports team, your city's streets—the list goes on. So we pave over big swathes of valuable real estate in our cities, creating asphalt heat islands that elevate urban temperatures and may contribute to climate change.

All of which is why Morgan Stanley financial analyst Adam Jonas calls the automobile "the world's most underutilized asset" and the auto industry "the most disruptable business on earth." It's why

Pulitzer Prize—winning journalist Edward Humes says, "In almost every way imaginable, the car, as it is deployed and used today, is insane."

— — —

I couldn't agree more. Thankfully, we've entered a period that is moving us toward a saner transportation solution—one of those rare disruptions that will improve the way life happens for decades, and perhaps for centuries, to come. This transformation will occur because it allows people to get around at lower cost more conveniently. Happily, the solution also happens to be better for the earth.

Many of the key players in the disruption turned to their work after a moment of extreme frustration with automobiles and the system they've spawned. For example, consider Google cofounder Larry Page, who, fatefully, did not have a car when he attended the University of Michigan as an undergraduate.

Page studied at the University of Michigan from 1991 to 1995 to get his bachelor's degree in computer engineering. He had a strong personal connection to the school; his grandfather, an autoworker for General Motors in Flint's Chevrolet plant, had driven Page's father and aunt around the Ann Arbor campus of the University of Michigan and told the children that they would one day attend the place. Both did. Page's father also met Page's mother there. So it was almost inevitable that Page himself would attend U of M.

Ann Arbor is a pleasant place in the spring, summer and fall, full of trees and rolling hills, student cyclists and joggers, the landscape dominated by the green of natural vegetation and the so-called maize and blue that are the school's official colors.

But in winter, the campus turns into a difficult place to be outside. Few people get around by bicycle between December and March because Michigan winters can be brutal. The landlocked campus is far from any temperature-buffering major body of water. Darkness falls by 5:00 P.M., and the cold is omnipresent. The sidewalks can feature slush and sleet in early winter, which then harden into black ice come January and February.

The other thing about Ann Arbor is that traffic can be terrible. In summer, it's bad. In winter, when the snowdrifts freeze into iron berms that narrow the already traffic-clogged roads, the car congestion and parking woes grow even worse. Those who don't have automobiles are forced to ride the bus, which arrives irregularly, and sometimes not at all.

Page would get out of one of his afternoon engineering courses and head to the bus stop and wait, shivering, looking down the road hoping to spy the distinctive headlight pattern of the buses used by the local transit authority. While automobiles passed by, their individual drivers embedded in little cocoons of warmth, Page would huddle in the shelter and hope for the arrival of a ride that never seemed to come, and he would think about how poorly we as a society had solved the transportation problem.

Consequently, Page became obsessed with alternative solutions. Those interminable minutes that Page spent waiting for the bus in the Michigan winter convinced him to draft, as a U of M student, an idea for a personal rapid-transportation system, an interconnected monorail on which two-person mobility pods were available on a moment's notice to ferry riders wherever they wished. Those frigid minutes also encouraged Page to join U of M's solar-car racing team—after all, cars that ran on free solar power would presumably make transportation more affordable to everyone. Finally, those minutes were a factor in Page considering to pursue, as a graduate student at Stanford in the late nineties, autonomous-car development rather than the world-changing search-engine project on which he eventually landed. And they spurred Page's interest in the desert and urban challenges that the U.S. military's Defense Advanced Research Projects Agency (DARPA) staged in California in 2004, 2005 and 2007. Those challenges led directly to the decision by Page and his partner, Sergey Brin, to fund Google's Chauffeur self-driving car project (now called Waymo), which convinced the world that autonomous vehicles were not just possible, but inevitable, and a lot sooner than many people expected.

My moment of greatest frustration with the old solution happened in Germany, where I was attending Frankfurt's 2001 International Motor Show. At the time, I was General Motors' corporate vice president of research, development and planning, and a member of CEO Rick Wagoner's thirteen-person strategy board, which was responsible for making the automaker's biggest decisions.

In Frankfurt, I was heading back to my hotel when my cell phone rang. It was GM security, which was unusual. What was even more unusual was the tension in the caller's voice. The security officer said he could not get into the details, but that as soon as I arrived at the hotel I was to proceed to a specific conference room.

I'd never received a call like that.

When I entered the conference room, several other GM Automotive Strategy Board members were present and the TV was turned on. I could see on the screen that one of the World Trade Center towers was on fire. Minutes later I watched a jetliner fly into the second tower.

It took three days before I was able to get home from Germany. I did a lot of thinking as those days passed. Many theories exist to explain why the attacks occurred. But it's impossible to ignore that one contributing factor was U.S. dependence on oil imported from the Middle East.

I couldn't help but feel as though the auto industry bore some blame for what happened. America was dependent on foreign oil because we needed it to power the cars and trucks that GM produced. Our customers enjoyed great freedom with GM products. But, I asked myself, was this freedom worth the price? For me, 9/11 screamed that the status quo of the auto industry, dominated as it was by gas-powered combustion engines, was unacceptable. And thanks to my job leading GM's R&D, I was in a position to do something about this. In fact, I felt like it was my *responsibility* to accelerate the development of alternatives to the current transportation system.

Soon, I developed a profile as the highest-ranked Detroit auto executive pulling for wholesale reform of America's automobile-based transportation system. (As I recall, the only other person in Detroit who was talking about the problems in the same way was William Clay Ford, Jr.)

Oil dependence, safety issues, traffic congestion and global warming—these and other ills were solvable, I argued in speeches and articles, if only we'd transform the auto industry. I focused on redefining the "design DNA" of automobiles based on electric drive and computerized controls, and I illustrated what was possible with the now renowned GM Autonomy concept car, which debuted at the 2002 North American International Auto Show in Detroit. (Autonomy was based on a skateboard-like platform similar to what underlies today's Tesla models.) I also steered GM toward a portfolio of alternative-propulsion systems based on hydrogen fuel cells, advanced batteries and biofuels, and arranged for GM to sponsor Carnegie Mellon's Team Tartan, which won the DARPA Urban Challenge by creating a robot version of a Chevy Tahoe. And as GM and its competitors fought to survive the 2008–2009 recession, I pushed to develop an autonomous, shareable and electric concept vehicle, the GM EN-V, that foresaw our self-driving future.

Those were the auto industry's darkest days, and while GM and Chrysler went bankrupt, and Ford mortgaged itself to narrowly avoid the same fate, a handful of auto industry outsiders began to challenge Detroit's dominance in a stunning convergence of new technology and innovative business models. This was the period in which Google gathered together the brightest engineering talent from the DARPA challenges and launched its Chauffeur self-driving car project. Upstart Tesla delivered its first Roadster in 2008, highlighting the promise of electric vehicles with outstanding performance using lithium-ion batteries. And shortly after that, scrappy start-ups Uber and Lyft, among others, established an enormous market for ride sharing and began the decoupling of people from personal ownership of automobiles. While Detroit was fighting for its life, the seeds of the mobility revolution were being planted by companies from outside the auto industry, by players with a bone-deep understanding of digital technology and a passion for designing and delivering compelling transportation experiences.

I left GM soon after the 2009 bankruptcy and, among other new positions, became the director of the Program on Sustainable Mobility at Columbia University, working out of economist Jeff Sachs's Earth Institute. There, I initiated the first research project to examine the economic implications of a future that saw transportation disrupted by three separate but related factors—shared-use vehicles, powered by electric motors, and driven autonomously. While each individual factor promised significant change, I was more interested in what would result when they converged. The expert math modeler Bill Jordan and I calculated in 2011 that the deployment of such an integrated system could reduce the annual costs of automobile travel in the U.S. alone by $4 trillion—about the same amount as the entire budget of the federal government. More to the point, our research suggested that driverless electric vehicles tailor-designed for shared transportation service in U.S. cities could reduce the out-of-pocket and time costs of conventional automobile travel by more than 80 percent (from $1.50 per mile to $0.25 per mile)—while providing safer and more convenient mobility.

Soon after I began that work I was recruited by Chauffeur's project leader, Sebastian Thrun, and engineering lead, Chris Urmson, as an adviser, a role I continue to hold today. In my eighth year at what is now called Waymo, advising one of the most exciting endeavors in engineering history, I feel lucky to have had the opportunity to work with Sebastian, Chris and such fascinating characters as Anthony Levandowski, Bryan Salesky, Mike Montemerlo, Dmitri Dolgov and Adam Frost, as well as Waymo CEO John Krafcik.

In 2018, Waymo achieved the realization of a dream that first gathered the team together in 2009—the deployment of autonomous, shared, electric vehicles. And the number of major companies testing these vehicles everywhere from Miami to San Francisco to

New York City is now approaching the dozens. Self-driving cars equipped with electric motors and deployed in a transportation-service model are poised to become the biggest thing to hit the automobile industry since the invention of the automobile itself. We are entering a new age of automobility, which redefines the freedom provided by today's automobiles, promising better mobility for more people at lower cost. The implications are profound, not just in terms of how our lives will change, but also for the automobile industry and everything it touches.

The resultant disruption will transform the way we live, the way we get around and the way we do business. It will virtually eliminate automobile crashes, radically decreasing the number of deaths they cause every year. It will decrease the cost of long-haul trucking by about 50 percent—a remarkable productivity-improvement opportunity and amplifier of e-commerce growth, and a profoundly upsetting prospect for the millions of employees and small-business owners who earn their living as drivers. The financial implications are compelling for the auto manufacturers, who will transition their business models from selling millions of vehicles to millions of different customers and instead operate massive fleets of self-driving taxis in population centers around the world. Today, the average net income per vehicle sold by most auto companies ranges from $1,000 to $5,000. In contrast, a transportation service vehicle with, for example, a 300,000-mile wear cycle earning just $0.10 per mile makes a lifetime profit of $30,000. (The 300,000-mile figure is based on the approximate lifetime of taxicabs with internal combustion and hybrid electric engines.)

This book is the story of the loosely connected visionaries who saw something was possible before others, how their visions have come to be and how this future will reshape our world. For their optimism, these few spent years being disparaged as futurists, as impractical dreamers, as kids playing in a sandbox—until suddenly, in the fall of 2015 and the spring of 2016, the industry recognized that the future the visionaries described wasn't just possible. It was practical and desirable, and coming sooner than anyone might have ever thought.

How these men and women pulled off that transition is a remarkable story—one filled with complex alliances and betrayals. It includes miracles of engineering and accidents of mechanics. Remarkable feats of software programming and quite a few questionable acts. Great sacrifice is made, as well as, eventually, wealth. There are heroes and villains, and a lot of characters residing somewhere in between.

The tale could feature many beginnings. You could say that it began at the 1939 World's Fair, where the General Motors pavilion provided a prescient version of a world much like the one we're approaching. I hope at least part of it began when I became head of GM's research and development, and CEO Rick Wagoner challenged me to reinvent the automobile. You could set the start of the sharing chapter near Boston, where Robin Chase cofounded Zipcar. The electric vehicle aspect started in Palo Alto, California, where Martin Eberhard and Marc Tarpenning, fresh off successfully selling one start-up, decided the new lithium-ion batteries deserved a shot in an automobile—and brought in an investor named Elon Musk.

But ultimately it was the autonomous end of this disruptive trinity that kick-started the transformation. Maybe that started coming true with the terrorist attacks of September 11, 2001, which in turn triggered a series of wars that spurred an obscure arm of the U.S. government, DARPA, to organize the challenges that ultimately set these dominoes in motion. But I'm not going to start this story at DARPA's home in Arlington County, Virginia. Rather, I'll start with the engineering student who probably sacrificed the most out of everyone—and who, going on fifteen years later, may turn out to be one of the ones who have gained the most, as well.

This story is going to start with Chris Urmson.

I

THE TURNING POINT

Chapter One

DARPA'S GRAND CHALLENGE

An engineer is someone who likes to work with numbers but doesn't have the personality to be an accountant.

—UNKNOWN

Over the last fifteen years of development on autonomous vehicles, if there is one figure who has been there, on the ground, getting his palms filthy with engine grease, breathing carbon monoxide exhaust and burning himself with electronic solder to solve each little problem as it comes up, it is Chris Urmson. The technical lead of the Carnegie Mellon University teams that competed in the three robot-car challenges staged by DARPA, Urmson's also the figure anointed as leader by the founder of Google's Chauffeur self-driving car project, Sebastian Thrun. In fact, Urmson ran day-to-day operations on the team from its founding in 2009 to shortly before the spinning out of Chauffeur from Alphabet into a stand-alone company, known as Waymo, in 2016. Finally, Urmson also played a key role in the power struggle that dominated Chauffeur for long periods of its existence.

To make this thing work, Urmson has sweated blood.

He'd be the first to admit he doesn't have the outright incandescent charisma of some of the other figures in this story. Urmson is smart, sure. He refined his willingness to consider every possible solution to a problem, no matter how outlandish, in the creative-thinking challenges that dominated the Canadian educational system's classes for gifted learners. What Urmson lacks is the bumblebee attention span of some of his self-driving colleagues. Perhaps this is because of the milieu in which he was raised. The oldest son of a prison warden and his nurse wife, Urmson grew up in small Canadian cities—Trenton, in eastern Ontario, where the biggest employer is a military base. Victoria, the seat of the British Columbia provincial government. The not-exactly-bustling metropolis of Winnipeg, Manitoba. His dad was rising through the ranks of the northern nation's correctional services bureaucracy, eventually running not just one prison, but a whole area of them, until the family settled in the sleepiest city of them all—Saskatoon, the capital of Saskatchewan, the least assuming province in one of the least assuming countries of the world.

Urmson grew up among people who viewed with suspicion those who drew attention to themselves. What the guy is, is solid. Straight-shooting. Steady. Urmson is not the guy you're going to notice first when you walk into a room. But you spend enough time with the people in that room, and I don't care who is in there, after a while Urmson will be the guy you trust to lead—to carry out the plan.

And in April 2003, Chris Urmson had a plan. In fact, Urmson thought he had the next couple of years of his life pretty well figured out as he drove from the remote Chilean city of Iquique to the great salt plains of the Atacama Desert. The road from Iquique into the Atacama would make anyone nervous. It zigs and zags from the Pacific Ocean on up a near-vertical shelf. Those who remember plate tectonics from high school geology might recall that this is where the Pacific's Nazca Plate collides with South America, pushing the continent into the air, creating a ridge thousands of feet high and a rain shadow that runs six hundred miles up and down the Chilean coast. That rain shadow is the Atacama. One of Earth's most forbidding landscapes, the driest nonpolar desert on the planet, an area so desolate that scientists use it as a stand-in for Mars. That was what Urmson was doing there. He was one of a handful of roboticists joining a team of NASA staff members to test a robot designed to crawl across the Martian landscape to seek out signs of life.

At twenty-seven, Urmson was tall and athletically built, with sandy blond hair and smiling blue eyes behind round, wire-framed spectacles. He tended to jam a baseball hat so low over his brow that the brim would touch the top of his glasses. Urmson planned to spend about a month in the Atacama. Then he'd return to Pittsburgh, where he was a graduate student in the robotics program at Carnegie Mellon University. He'd write up his dissertation, undergo the grilling every thesis committee is supposed to do, hopefully get his PhD and then a job—maybe join the faculty of his alma mater's Robotics Institute, home to more robotics brainpower than anywhere on earth, or maybe join one of the start-ups that occasionally spun out of the university. In any event, he'd start making money, enough for him and his wife to have the kids they'd been putting off while Urmson finished his studies.

The campsite that Urmson's research group chose amounted to little more than a handful of bright yellow dome tents, a slightly bigger meeting tent—where they kept the computers—and a pickup. And Hyperion. Hyperion was the robot. Not the conventional kind of robot. No arms and legs. Rather, Hyperion sat on a quartet of bike tires and was roofed with solar panels and powered with an electric motor. Hyperion was the reason why Urmson, and his fellow scientists from Carnegie Mellon and NASA's Ames Research Center, had traveled over half the planet.

Hyperion was designed to wander across the Martian surface, sniffing and scraping and testing the soil for signs of life. Urmson was in charge of programming the software that dictated how fast Hyperion rolled.

The scientists took their breakfasts and dinners at a nearby salt mine. Nights, they sat around a fire and watched the camanchacas roll in, the Pacific salt fog that could rust exposed metal within a single night. They turned in to tents they used for warmth rather than protection. You camp in other deserts, you need a tent to keep snakes out of your sleeping bag and scorpions out of your boots. But nothing lived in the Atacama Desert. Not snakes. Not scorpions. The only living things that Hyperion's minders saw were vultures.

The encounter that would change Urmson's life started with the sight of a long dust cloud led by a speeding pickup. Some minutes later, the dust cloud followed the pickup into the Hyperion campsite. The door opened, and out of the truck popped William L. Whittaker, commonly referred to as Red.

Whittaker was another big guy, an inch or two taller than Urmson at about six-foot-three, with shoulders that look like they'd brush the sides of interior door openings. His scalp is closely shorn; years ago, when he did have hair, the color of it was what gave him his nickname. His gaze is intelligent and contemplative. It feels like his eyes can see right into your soul when he looks at you. Anyone who spends five minutes with Whittaker can tell that he spent formative years in the U.S. Marine Corps. He speaks in the sort of aphorisms that drill sergeants put on their bedroom walls. "Winning isn't everything," he might say. "It's the *only* thing." And: "Worry is a formula for failure." Another favorite: "If you haven't done *everything*, you haven't done a thing." Hyperion was somewhere around the sixty-fifth robot Whittaker had worked on in his career as a roboticist.

The Carnegie Mellon professor strode out of the pickup in boots, conducting a round of handshakes with his big hands. He was there in part because he was Urmson's thesis adviser, and he was checking on his charge. But you could tell that Whittaker was holding something in. Something big. Pretty soon, Whittaker came out with it. The U.S. Department of Defense was staging a driving race for robots. Specifically, the Defense Advanced Research Projects Agency. DARPA, Urmson knew, was the U.S. government's developmental laboratory, credited with spurring such useful inventions as drone technology and the Internet (a military invention whose distributed knowledge network was intended to safeguard the records of the U.S. government in the event of a nuclear attack). DARPA was also responsible for such less-than-useful innovations as mechanical lobsters for the U.S. Navy and DNA-editing techniques intended to create humans who didn't need sleep. Now DARPA Director Tony Tether was turning the agency's direction toward autonomous cars.

For years, Washington had pushed American defense contractors to develop autonomous technology so that a third of all American military vehicles could be self-driving by 2015—a stated mandate from Congress. In the aftermath of 9/11, the effort took on added urgency as the U.S. military lost infantrymen and -women to improvised explosive devices planted under the roads in Afghanistan and Iraq. If self-driving vehicles ever became possible, military robots might drive themselves over the sort of desert roads found in overseas theaters of war. But the four-star generals had been frustrated with the pace of change. The problem was proving too difficult for the military contractors. And so Tether struck upon a novel solution: DARPA would stage a race. For robot cars.

As Whittaker recounted the details to Urmson, they sounded a little insane. DARPA said it would allow any American team to enter—student, hobbyist, professional, whomever. The course would bisect the Mojave Desert, running eastward from Barstow, California, to Primm, Nevada, for a distance of about 150 miles. The prize money would go to the first team that could do it in under ten hours.

"Wow," Urmson said, thinking Whittaker was just making conversation.

But Whittaker never just made conversation. The prize money, the old marine said, was a million bucks. And Whittaker wanted to win that money with Urmson's help.

— — —

It would be three years before I met Chris Urmson, who would go on to become one of my favorite people. But I can see how this situation would have presented him with a dilemma that contradicted two of his prime directives. Urmson had a seemingly innate desire to try to improve the stupid and inefficient things about the world; he once interrupted an important business meeting at a Pittsburgh coffee shop to burst out onto the street and direct traffic, just to help someone turn left out of a parking lot. He was programmed with an engineer's duty to seek out the coolest and most interesting projects that could change the lives of the most people. Which is why Hyperion was such a perfect project for him. How could you get cooler than an autonomous robot designed to seek out life on other planets?

Actually, it turned out that you could. Urmson's work with Hyperion was helping the robot travel anywhere from 15 to 25 centimeters a second—about the pace of a slow walk. In the DARPA race, the robot would have to travel 150 miles in at least 10 hours, which required an average speed of about 15 mph, as fast as most cyclists went. The speed, the money, the fact that the race was intended to address an issue killing American soldiers overseas—Urmson got it. He ached to participate.

But there was a problem: He was also programmed with a duty passed down to him from his parents, to do what was best for his family.

Chris Urmson was born in 1976 to Paul and Susan Urmson, an English couple who had immigrated to Canada because they thought it would represent better opportunities for their three sons. Paul's first career was as an electrician, and then, once his kids were born, he pursued his college degree at night school, earning his BA and then master's. Susan enrolled in nursing school after the kids were born and went on to administer methadone programs within the Canadian prison system.

The point? The three Urmson boys grew up in homes where the parents were always working, always bettering themselves for the sake of the family and where education was prized from the kids' earliest ages. The Urmson parents ran their lives for their children. The family moved a lot because Paul's work in the prison system required him to transfer around the country. Each time they did, Paul and Susan settled the family in the cheapest house in the nicest neighborhood they could find—a strategy they devised to send their kids to the best public schools. The strategy worked. In addition to birthing one of the most important engineers in the development of autonomous cars, the Urmsons also raised an orthopedic surgeon and a Mountie, a member of the Royal Canadian Mounted Police, which is something of a trifecta for middle-class families north of the border.

At a young age, Chris's teachers assessed him as gifted, which qualified him to attend special classes with similarly intelligent children. The classes provided the ability to conduct independent projects. Gifted-program teachers encouraged their students to enter a series of science fairs then known as Olympics of the Mind, which challenged participants to solve unconventional problems. How do you build a tower out of just paper towel tubes? Propel a toy car with a mousetrap? Safeguard an egg dropped from an extreme height?

The experiences set Urmson up well to compete in Canada-wide science fairs. The year the Urmson family moved from Victoria to Trenton, the national finals happened to be held in Victoria. Urmson ached to visit his old friends, and so he directed all his energies toward winning the local competition. His entry, "Striking News About Impacts," predicted the direction a body would travel after a collision. He won the Trenton fair, and received the free trip to Victoria.

Bit by the science bug, Urmson followed up with a project involving a model of ionic propulsion—"Ionic, Isn't It?" was the project name. It not only won him another trip to the Canada-wide competition, but also garnered him second prize. Another year he won a silver medal at the national level and qualified for a four-week trip to study programming at Israel's Weizmann Institute. Urmson would go on to study computer engineering at the University of Manitoba, where one of his projects entailed building a robot that traveled autonomously around a darkened room, seeking out the brightest sources of light.

Urmson was torn in his last year of university. One path, favored by mothers everywhere, might have seen Urmson going on to med school. Except that didn't exercise his yen for building things, for envisioning complex systems and then figuring out how to make them work. Wandering by the office of his computer engineering department one day, Urmson's eye was caught by a remarkable poster: a vehicle, maybe some sort of a planetary rover, climbing up and out of some sort of crater. "Come be a part of the robot revolution!" the poster read, with information about attending Carnegie Mellon University. It was a career based on the sort of thing Urmson had been doing all his life. Olympics of the Mind. Science-fair stuff. He applied, and ended up in Pittsburgh the following year.

At Carnegie Mellon, Urmson met Red Whittaker, who by 2003 already was a legend in American robotics and one of the best-known robot designers in the world. Born in 1948, Whittaker was fifty-five in 2003 and had become widely known for his willingness to take on projects that everyone else thought impossible. "If there is anyone in the world who can find a way to make things happen, it's Red Whittaker," said one colleague.

Whittaker may have been genetically programmed to ignore the impossible. His father was an air force bombardier in World War II who would go on to sell explosives to mining companies. His science-teacher mother was an amateur pilot who once flew under a bridge while the young Whittaker was in the plane alongside her. After serving in the marines for two years, Whittaker attended Princeton University, earning his degree in civil engineering in 1973, and then attending graduate school at Carnegie Mellon.

Whittaker made his name after the partial meltdown in 1979 of the Three Mile Island nuclear-generating station, America's worst-ever nuclear accident. Cleaning up the incident required getting into the reactor's basement to learn how radioactive the site was. Several contractors spent almost a billion dollars on the cleanup but still couldn't figure out how to get inside. When Whittaker asked for his shot, the government figured they didn't have anything to lose. Whittaker reasoned that, while the radioactivity prevented humans from getting into the reactor, machines should have no problem. He created a three-wheeled Remote Reconnaissance Vehicle, known as "Rover," which he operated by remote control. Rover successfully made it to the basement. Best of all? The program cost only $1.5 million, which the government considered cheap.

Since then, Whittaker had specialized in building robots designed to work in harsh environments. One of his creations explored volcanic craters. Another mantis-like contraption built structures in space. Still another, created with a team that included the German software wizard Sebastian Thrun, crawled through the darkness of long-abandoned mines, mapping their interior passageways. Urmson had worked with Whittaker to develop computer algorithms designed to increase the speed at which robots were able to travel autonomously.

When Urmson returned from the Atacama Desert, he had a tough conversation with his wife, Jennifer. Urmson wanted to set aside the completion of his PhD for a time and pursue the DARPA race with Whittaker. The DARPA Grand Challenge was the talk of their academic specialty. DARPA had figured it might be lucky to get twenty entries. Eventually, 106 teams would enter. Urmson felt he had no choice but to join. Who knew what sort of fascinating epiphanies would emerge from the project? Who knew what Urmson would miss if he didn't take part?

Urmson convinced Jennifer to let him do one race. The couple would put off having kids until Urmson was done. But then fate threw them a slider: It turned out Jennifer was already pregnant. The news added to the pressure Urmson felt to win. After all, it was the best way to ensure he'd get a high-paying job, once he finished.

To attract a team, Red Whittaker put up posters all around the Carnegie Mellon campus advertising an unconventional, graduate-level seminar class, Mobile Robot Development. It was pass or fail, and it featured only one assignment: to build a robot that would win the first DARPA Grand Challenge. He also sent out email blasts to potential sponsors and volunteers, which featured his trademark bravado: "The race defies prevailing technology, and many hold that the challenge prize is unwinnable in our time."

Whittaker staged the first meeting of the team in a Carnegie Mellon seminar room on April 30, 2003, according to Wayt Gibbs, a reporter the magazine *Scientific American* embedded in Pittsburgh. "Welcome to the first meeting of the Red Team," Whittaker began. "I am committed to leading this team to victory in Las Vegas next year."

The men and women in the room were about as motley a crew as it was possible to put together in Pittsburgh tech circles. Bob Bittner was a former combat engineer who'd spent six years in a submarine. Spencer Spiker was a retired helicopter test pilot, a West Point–educated mechanical engineer who led two hundred people as a company commander in the U.S. Army, and who had left the service to spend more time with his family—then found himself jobless in a severe recession. He joined Red's team because he had nothing better to do, then worked himself into a full-time staff position. Michael Clark was a NASA engineer who used a wheelchair to get around; having fallen on hard times, he lived for a spell out of his van. Lots of people had seen Red's poster, apparently, and lots of people were inspired to work on the project it advertised. "I don't know anything about computers—but I'd like to volunteer," said Mickey Struthers, a postman who showed up to the first class because he wanted to participate in a historic science project.

"You've got a warm body"—Whittaker grinned, shaking Mickey's hand—"and we need warm bodies."

They began their efforts brainstorming what sort of a vehicle they would use. DARPA had announced the course would be designed by Sal Fish, the operator of such tough off-road races as the Baja 1000. Red Team figured it would have to be prepared for a course that would wind through dry river gulches, box canyons, mountain ridges, rocks, sagebrush and cliffs. So the robot they designed had to be able to either get around such land features or drive over them.

No idea was too outrageous to be considered. One of the first suggestions was a giant tricycle that had wheels seven feet in diameter. The team discussed using a Chenoweth combat dune buggy, a low-slung contraption on four fat tires favored by mercenaries and warlords. Other brainstorming options included construction equipment, an all-terrain vehicle and a tank. But the team ultimately opted for pragmatism. After all, Whittaker figured the budget to develop the robot would be around $3.5 million. Labor aside, $725,000 of that entailed the cost of the products required to build the vehicle. Whittaker was crisscrossing the country to find sponsors. Intel, Boeing and Caterpillar all kicked in some money. Google, which everyone thought of as a search engine company at this point, sponsored Red Team to the tune of $100,000 after Whittaker visited its headquarters in Mountain View, California, and met both Larry Page and Sergey Brin. But such funds wouldn't go far when you were trying to build the fastest-ever robot car. Red had bought a cattle ranch a couple of hours east of Pittsburgh in the early nineties because he felt his academic life was too sedentary and sought physical activity that worked his muscles, not his mind. In September 2003, with the March 2004 race date fast approaching, Whittaker finally bought the vehicle that would become their robot from another farmer in the area.

Some in the class were astonished when they saw it. Shouldn't a self-driving car look cool, and polished and, um, high-tech? The vehicle Red had procured was the opposite of high-tech. It was a High Mobility Multipurpose Wheeled Vehicle M998: a Humvee, battered by time. It was seventeen years old. No one had any idea how many miles it had, because the vehicle didn't have an odometer. Nevertheless, the price was right: $18,000. The *key* thing was, it worked.

Whittaker was under a lot of pressure. Around the country, dozens of robotics enthusiasts were working to create entries for the challenge; so many, in fact, that DARPA was requiring everyone to submit a detailed and academically rigorous declaration of the approach they planned to take. The step was intended to limit the race entrants to serious competitors. There were high school students and bored mid-career engineers. Several were former contestants on the mechanical-gladiator game show, *BattleBots*, which featured remote-controlled robots fighting to the death, or at least, deactivation. Regardless of where they came from, the competitors all seemed to have one goal in mind: Beat Red Whittaker's team. Why was the CMU team in so many other people's sights? Whittaker's team was the biggest, with thirty members. It was one of the best funded. And many also believed that it was DARPA's hoped-for winner.

Red's leadership style was to take a bunch of people, introduce the problem to them, set ambitious and clearly defined goals that reflected progress toward the solution—and then get out of the way. He'd drop in regularly to check in and apply pressure on his charges. Such visits could be intense. According to a *Wired* article, Whittaker once drew an analogy between developing robots and the labor required to construct the enormous historic monuments around the Nile. "If you're in Egypt building the pyramids, you've got to have slaves," he said. The implication? Whittaker's students were his slaves. One of Red's longtime students, Kevin Peterson, who would become Red Team's software lead, had attended Princeton High School, where he encountered Dr. Anthony Biancosino, the domineering music teacher on whom Damien Chazelle loosely based the bandleader in the 2015 movie *Whiplash*. Peterson responded to Whittaker's style, he says, in part because he'd already been through his experiences at Princeton with "Dr. B." "There was an ethos around both of them of being larger than life and somewhat mysterious," Peterson recalled. "The idea that you need to work hard to be part of their exclusive team if you want to join them. They're both up to big things and you need to be a badass to be on the team. Funny thing is, both of them would accept and build anyone who had that level of dedication. It's more about hard work than initial skill." One of Whittaker's favorite motivational anecdotes placed his charges in the roles of the Inuit in the Arctic, who had to decide which strategy to use when seeking out food. Are you going to go out and try to find a few berries and bits of lichen? Whittaker would ask his team. Or are you going to find and kill the walrus that feeds the whole village?

Sometimes it was hard to tell what Whittaker meant by his stories. Peterson interpreted this one as a challenge. Were you going to go about your life just getting by? Or were you the type who was going to go out and give your best effort to do something awesome?

Realizing that his course would require more work than they were prepared to give, some people dropped Whittaker's class. The ones who remained essentially dropped every other one of their classes and just worked for him. Peterson was one of the ones who remained. He gave up his social life, as well as communicating with his family. He even gave up sleeping. Several months in, he became so sleep-deprived that he fainted. The problem was that he was going down a set of stairs when he did. He hit his head, was taken to the hospital to be assessed—and was back working on the project within a few days.

Empowering inexperienced and sleep-deprived graduate students who were totally committed to the project's success could create some unusual situations. One morning, Whittaker and Urmson arrived to check in on the students and volunteers and were met with the results of one of these hyper-caffeinated work sessions: Their treasured Humvee no longer had a roof. Working through the night, one of the student team members had decided that the Humvee's interior didn't have enough room to store the batteries and computers and actuators that the self-driving equipment would require. So he went and got a Sawzall and cut through each one of the Humvee's roof pillars, essentially decapitating the vehicle.

This was the sort of initiative that would typically have been applauded by Whittaker. Except the impromptu roof amputation wasn't really necessary. Even if the equipment couldn't fit in the Humvee's cab, they could have ripped out some seats, or mounted additional equipment *on* the Humvee's roof. Removing it made the vehicle illegal to drive on public roads. From then on, whenever they wanted to take the Humvee to the sort of wide-open space where they could test it, they would have to tow the vehicle—an ignoble start for a robot that was supposed to drive itself.

To provide the Humvee with the ability to drive itself, Red Team essentially reverse-engineered the sensory tools humans use to help them drive. The vehicle needed, for example, eyes to see—and so the Red Team procured several types of LIDAR (Light Detection and Ranging) devices. The LIDAR's job was to shoot out beams of light and sense when the beams bounced back. Precisely calculating the timing of the beams' return allowed the LIDAR to determine how close the sensor was to the object that the light beam bounced against. Repeated thousands of times per second, the LIDAR could create a rudimentary picture of the world outside the vehicle.

The main LIDAR sensor would allow the robot to detect obstacles seventy-five meters ahead. Three supplemental LIDAR devices scanned a wider field of view within twenty-five meters of the robot's front end. A stereo-vision processing system represented a different way to use light to detect objects, employing a pair of cameras. But the cameras and LIDAR might have trouble penetrating the dust clouds that could arise on sandy desert roads. To provide a sense of the world in dusty conditions, Red Team also bought a radar system that used sound to detect obstacles.

To control the vehicle's direction and speed, Red Team wouldn't be able to use a foot on the gas pedal or a hand on the steering wheel. Actuators would take their place. Essentially, these were electric motors that twisted, pushed or pulled—to make the vehicle accelerate, brake or turn left or right.

Sitting in the center of all that was a series of computers, the robot's brain. Donated by Intel, one was a quad-processor Itanium 2 server that featured three gigabytes of RAM. Some of the computers were intended to combine the information provided by the LIDAR, the stereo-vision system and the radar sensor to create a model of the world. Another computer employed GPS data and motion-tracking tools to locate the robot in the world within a single meter of accuracy. Now that it had a conception of its surroundings and knew its location, the robot's computer system would have just two questions to answer. Two questions that humans asked themselves, thousands of times a trip: How fast should I be going? And where should I be steering?

Whittaker scheduled one hundred days to actually get the robot assembled and the software built. The deadline fell in November, but as Thanksgiving approached, significant portions of the vehicle remained unfinished. The computers weren't wired together, for example. Nor were the sensors mounted. The robot did have a name, though: Sandstorm, after the dust clouds the vehicle would kick up in the Mojave.

Whittaker and Urmson both worried a lot about the Mojave Desert. They worried about the off-road conditions of the course, and the effects of the Mojave's rutted roads on their sensitive sensors and microprocessors. Driven over even at moderate speed, the Mojave Desert's rocks and ridges were bound to create vibrations that the students believed had the potential to damage the computer's memory. After all, your basic disk drive is just a magnetic metal plate that spins really quickly. They're encoded by a precise bit of metal that hovers just above the plate. Extreme shocks could see the metal stick gouging chunks from the spinning plate and damaging the drive. Those same bumps could create false readings from the sensors.

Consequently, Red Team spent a lot of time determining how to insulate the computers and sensors from the jars and bumps that would happen as the Humvee drove across the desert. The solution, they decided, was to protect the equipment the same way automobile manufacturers insulated humans from bumps and jars. With springs and shock absorbers, which were fitted to an enormous metal box where the Humvee's roof used to be. Dubbed the "e-box," for electronics box, the 1,200-pound container didn't just contain hard drives. It also encompassed much of the robot's most sensitive equipment—the computers, the GPS system, the radar as well as the supplementary LIDAR units.

The main LIDAR and the stereo-vision device still remained

sensitive to the pitches and rolls that could strike the robot as it navigated the off-road trail. So the team spent untold hours engineering a device based on old nautical gimbals, complex series of interconnected arms and pivots that kept a ship's compass stable in even the heaviest of seas. Part of the Red Team designed and built their own gimbal, mounting inside it the main LIDAR and the stereo-vision system, and protecting it all in a sphere a little larger than a classroom globe. Little motors in the gimbal allowed Sandstorm to direct the LIDAR and camera wherever it needed to sense the world. Heading into what its onboard map told it was a leftward curve, the LIDAR would "look" to the left so that it could see the world in the direction of the world to come.

As technical director, Urmson was the one in charge of putting all these pieces together. He felt enormous pressure both at home and on the Red Team. That September, his wife had just had the couple's first child, a baby boy. But Urmson couldn't spend much time at home. He had made a promise to Whittaker that the robot would drive itself the entire length of the race, 150 miles, by midnight on December 10, 2003—three months before race date.

To meet that deadline he was working sixteen-hour days, seven days a week; during one furious round of assembly Urmson didn't sleep for forty hours. The week before Thanksgiving, Whittaker added to the pressure. "This vehicle hasn't rolled so much as a foot under its own control," he said during one meeting with Urmson and other key team members, according to the journalist Wayt Gibbs. "You have promised to get 150 miles on that beast in two weeks . . . Anyone who thinks it is not appropriate for us to go for 150 miles by December 10, raise your hand." Silence. Not a single person elevated an arm. Whittaker smiled, according to Gibbs, and made an observation in his characteristically florid language: "We're now heading into that violent and wretched time of birthing this machine and launching it on its maiden voyage."

The assembly work happened in a big garage in Carnegie Mellon's Planetary Robotics building. Envision the best mechanics shop you've ever seen, and you'll be close to this workspace. The ceiling is a few stories tall, with gangways and a small-scale version of a crane, the better to lift heavy objects. Lathes and drill presses, drawers full of every implement imaginable, as well as computer diagnostic equipment—every available horizontal surface features tools. It is the kind of place where you could literally make almost anything.

The venue would host Urmson and the members of his team pretty much nonstop through that Thanksgiving weekend. By the end of it, enough computers were wired together, and enough sensors mounted, that Sandstorm felt like it was coming alive. It was around this period that the team found the perfect place to test their Frankenstein's monster. There weren't many spots with convenient access to the CMU campus where a 5,000-pound, exhaust-snorting, diesel-gulping, oil-dripping robot could push the limits of its abilities without risking civilian fatalities. It was Mickey Struthers, the postman volunteer, who thought of the solution. One day while he was driving over Pittsburgh's Hot Metal Bridge on the way to Carnegie Mellon, Mickey noticed the lights along the shores of the Monongahela River twinkling in the cool evening air. All except for a vast swathe of dark shoreline to the right of the bridge. Mickey knew that was industrial land that had once housed Pittsburgh's last steel mill, the LTV Coke Works, which had closed in 1998. Since then the land had sat fallow.

Struthers suggested the site to Whittaker, who loved the idea for both its convenience as well as its industrial heritage. The 168-acre land parcel housed a railroad roundhouse and numerous outbuildings and equipment that made it seem as though it was left over from the industrial revolution, connecting the team to the same brawny spirit that had built Pittsburgh so many decades ago. With a few phone calls to the wealthy family foundations that owned the land, Whittaker arranged for the team to test there.

On the second of December, the team took the first of what would become many test runs at the Coke Works. The distressed location with its spent oil cans and rusted industrial detritus seemed appropriate for the ancient-looking Humvee, which just in general seemed to have more in common with a Jurassic-era dinosaur than one of the most innovative mechanical devices ever assembled. Snow covered the ground. The temperature was eighteen degrees. "Just like the Mojave Desert, huh?" shouted one team member, according to a *Wired* article. (Whittaker, meanwhile, was wandering around in a knit shirt, jeans and boots he wore without socks.) Urmson climbed aboard for the first run to manually hit the emergency stop button if the robot suddenly went crazy. The robot swerved toward a precipice when first activated, then settled and drove its course as expected. After a few uneventful laps Urmson decided, at 7:51 P.M., to see what would happen when he gave Sandstorm free rein. He clambered off the robot. The team programmed in a series of GPS waypoints that drew a dot-to-dot version of an oval. Not sure whether to breathe, the team watched the robot roll along its route for half an hour, ultimately accumulating four miles. No accidents. No incidents of any kind, in fact. They were nowhere near making their 150 miles yet, but that evening, it was difficult to deny they were progressing toward their goal.

Another week passed, and late in the evening on the tenth of December, with just a couple of hours before the midnight deadline by which Urmson and the team had promised Whittaker that Sandstorm would be able to drive 150 miles on its own, the robot was not cooperating. Bugs arose in the self-driving software every time it drove more than a few laps. Urmson and his fellow teammates had been camped out for days at the Coke Works, if you called camping sleeping in your running car with the heat on full blast. Despite daylong debugging sessions, Sandstorm remained unpredictable and occasionally suicidal—lurching into a telephone pole, catching fire, becoming suddenly unable to sense GPS signals. A calm spell saw the Humvee revolving the track, again, again, again, and then for no apparent reason, swerving off course and running itself through a chain-link fence before Urmson could activate the e-stop. Sometime later, with Sandstorm liberated from the barbed wire and the deadline approaching, Whittaker gathered Urmson and everyone else around him, according to Gibbs. Sure, the December 10 deadline approached—but even if it passed, Whittaker vowed, they'd continue their work, through tomorrow, and even the next day if necessary, until Sandstorm achieved the 150-mile goal. "We say what we'll do, and we do what we say," vowed Red in *Scientific American*.

Then it started to rain—a frigid December drizzle that soaked clothing and chilled to the bone. Sandstorm was not well protected against rain. One of the dozen or so team members still on site spread a tarp over the robot's computer equipment. Red wasn't around. Gibbs wrote that Urmson looked at his teammates, shivering in dripping lean-tos under blankets. He thought about the possibility of the falling moisture disabling one of their sensors, or shorting out a processor. Perhaps he also thought about his wife and baby boy back home. And he decided to send the team home.

Whittaker was livid when everyone showed up to the Coke Works the following day, Gibbs reported, comparing the team leader to "an angry coach at halftime." He ranted about all the sacrifices they'd made to try to achieve the 150-mile goal. The shop was a mess, the robot unpainted, the website out of date—all that work went undone as everyone concentrated on getting Sandstorm in the sort of shape required to make its race run. To a roomful of people unwilling to meet his gaze, Whittaker said, "Yesterday we lost that sense deep inside of what we're all about. What we have just been through was a dress rehearsal of race day. This is exactly what the 13th of March will be like. We're in basic training; this is all about cranking it up a notch. Come March, we will be the machine." Whittaker concluded his venting, Gibbs reported, by asking who was willing to work all day, every day, for the next four days, until they completed their nonstop 150-mile run. Fourteen team members in the room raised their hands. Including Urmson.

Two days later, U.S. soldiers captured Saddam Hussein in a spider hole near Tikrit, and the war in Iraq dominated headlines and the cable news channels as it never had. Every day, the news seemed to feature more casualties from IEDs in Iraq or Afghanistan—fatalities Red Team members hoped the robot vehicles might one day prevent. Then the overseas conflicts supplied Urmson with an idea.

In recent years, maps had become a crucial component of successful robotics. Maps allowed robots to locate themselves in the world much more accurately than GPS alone. A technique called simultaneous localization and mapping, abbreviated to SLAM, saw a robot scan an area with LIDAR to map the permanent landmarks—in exterior spaces, things like trees, light poles, road curbs and buildings. Then, the next time the robot traveled the same territory, it would consult its map and compare its position relative to the previous landmarks, to get an ultra-accurate idea of where it was. Problem was, Sandstorm couldn't use this technique, because DARPA was keeping the race location secret.

Then, one day, Urmson was watching coverage of the war on one of the cable news channels. The scene will be familiar to anyone who lived through the post-9/11 period—a grainy portrait of an SUV traveling fast along a remote desert road. From somewhere in the distance, a rocket blazes into the picture, collides with the SUV and obliterates the vehicle in a blast of dust and metal.

The footage of the successful deployment of a laser-guided bomb was captured by a camera-equipped drone aircraft. The drones flew above the conflicts to provide imagery of the Iraqi and Afghan territories. Drones were searching Afghanistan for Al-Qaeda hideouts that might shelter Osama bin Laden. They were scanning Iraq for nests of Ba'athist loyalists.

If the U.S. military could use drones to obtain imagery of places so hostile and remote, Urmson thought, then such imagery would soon be available for the entire world. And perhaps, Urmson reasoned, that same type of imagery could be used to simplify the robot's task. They weren't able to use LIDAR to scan the race course in advance, because no one on Red Team knew where the race course was, but they did know the race went across the Mojave Desert—and maps existed of that, didn't they? In fact, portraits of the Mojave had already been built by entities like the U.S. Geological Survey and the military.

"We realized we didn't have to do SLAM," Urmson recalled. "Because it was becoming clear there would be a global database [of maps] available . . . So why not use them?"

If Red Team members could give Sandstorm an accurate map of its surroundings *before* the race, they could remove a time-intensive step from the computational task. The new approach reframed the challenge. The team had assumed they were trying to build a robot that could sense the world so well, it could discern a road in the desert and navigate it safely for 150 miles. Using maps meant the robot could be told in advance where the road was, and how to drive it. The method had the potential to allow Sandstorm to travel much faster than it otherwise might.

But first, Red Team's undergraduates, pauper grad students and volunteers would have to build the most detailed map of the Mojave Desert ever assembled. It was an enormous task, but Red Whittaker's students were accustomed to achieving enormous tasks. A portion of the team set to procuring high-res maps of the whole of the Mojave Desert, a relatively simple matter, given Whittaker's and Spencer Spiker's defense contacts. Now the team set about using the maps to plot routes through the Mojave. They also dispatched two engineers, Tugrul Galatali and Josh Anhalt, to drive as many roads in the Mojave Desert as possible in a rented SUV with video cameras sticking out the windows, capturing imagery from the ground in what amounted to an early, rudimentary execution of Google's Street View idea.

The next step saw the Carnegie Mellon mapping team comparing the footage and the map to assign each area with a value—what they called a cost. So a ridge or a cliff that would wreck Sandstorm if the robot went over it would get a cost of infinity. A smooth road or a dry, flat lake bed likely would have a cost of zero. Sandstorm's computers then were programmed to direct the robot to drive the route with the lowest cost.

One evening, with just weeks to go before race date, the senior members of Red Team met in the loft of Carnegie Mellon's Planetary Robotics building. "We were making some progress, trying to map every trail in that whole desert," Urmson recalls. But at some point during this meeting in the loft, Urmson realized their work wasn't happening quickly enough. "It became clear we weren't going to get there," he said. Too many different potential routes existed. By the time the race date arrived, they would have mapped out only a small portion of the possible routes.

That was the point that Red Team came to its second epiphany. To reduce the possibility of exactly this sort of advance route planning, DARPA had told the teams that its staff would wait to disclose the precise course until just two hours before the start—at 4:30 A.M. the morning of the race. Red Team was getting good at creating routes through the desert. So what if they changed strategies? What if, rather than focusing on creating a map that featured a pre-driven route along every single conceivable trail through the desert, they instead became really good, and blindingly fast, at teaching Sandstorm to drive a *single* trail?

Rather than a perfect map, they thought, why didn't they focus on creating a single, perfect *route*? One they could plan out in the two-hour span between the time DARPA disclosed the approximate course and the start of the race? The old way involved using the maps and the route planners during the months before the race to effectively pre-drive every single road through a desert that covered a territory of fifty thousand square miles. This new way involved focusing on a single 150-mile path that the planning team would examine in fine detail—and doing it in the 120 minutes that passed after DARPA disclosed the race route.

From that moment on, one part of Red Team focused on executing the second epiphany. In the old high bay in the Planetary Robotics building, about a dozen members rehearsed exactly what would happen after DARPA handed over the route in a computer file at 4:30 A.M. The file would feature a series of about 2,500 GPS waypoints, which everyone referred to as "breadcrumbs," spaced about a hundred yards away from one another, tracing out the course in a dot-to-dot fashion. The dozen members of Red Team's planning unit would leap into action. One would feed the file into a software program that used the Mojave map's cost estimates to build a more precise route, with many times more breadcrumbs than DARPA's route network definition file (RNDF).

But Urmson, Whittaker and their team didn't trust the route calculated by the planning software. It had been known to send Sandstorm on journeys that went over ridges, into ditches or through wire fences. So a team of editors would divide up the course into sections and then, using computers, virtually go over every yard of the computer-calculated race path to make sure the software hadn't made any mistakes. Once the human editors were done correcting the course, they'd reassemble it into a single route and upload it to Sandstorm, to execute on the race course.

Still, by January 2004, just two months before race date, Sandstorm had not yet gone fifty miles on its own. One thing causing Whittaker and Urmson anxiety was the disconnect between where they were testing Sandstorm and the race course. They were testing the robot on the frigid shores of Pittsburgh's Monongahela River. The race would be held in the Mojave Desert. Would the change in environment pose a problem to Sandstorm?

In February, Whittaker arranged for some of the team's key members, including Urmson, Peterson and Spiker, to accompany Sandstorm to the Mojave Desert to refine the robot's capabilities. (Sandstorm actually made the trip in a fifty-two-foot enclosed semi-trailer.) The final part of preparations would happen at the Nevada Automotive Test Center, an enormous swathe of desert where companies from all parts of the automotive sector, from tire manufacturers to transmission firms, tested their products in the harshest desert terrain available.

In Nevada, Urmson's team worked exclusively on Sandstorm. Write code, take Sandstorm out to test the code, watch for mistakes, take note of the mistakes, write code. They repeated the cycle without regard to clocks or arbitrary separations of day and night.

Two, three days at a time they worked without sleeping, fueled by Mountain Dew, Red Bull and junk food, and then, when they were too exhausted to manage to keep themselves vertical, they slept. Sometimes in an RV they'd rented, although the trailer didn't have enough beds for all of them; others slept on the floor of the test center's mechanics shop on folding lawn chairs, or in the reclined seats of the SUVs they rented to tail Sandstorm.

Working nonstop, through night, through day, the way they did presented some difficulties. One evening, past midnight, Sandstorm ran into a fence post, wrecking the front bumper, which was necessary to support cameras and radar sensors. The test center's mechanics building was locked up, of course, but in the spirit of asking for forgiveness being easier than requesting advance permission, Spiker and one of the students scaled the fence and broke into the building, where they welded together an entirely new bumper with thick steel pipe. The thing ended up weighing about two hundred pounds—making it more than able to support the sensing equipment the robot required. "You could probably have driven through a building and not hurt that thing," Spiker recalls.

One thing they didn't do much of was bathe. The wastewater tank in their rented RV filled up, and by the time they got around to driving it to the nearest town to empty it, the vibrations from the washboard dirt road into town splashed sewage all over the RV's interior. Cleaning the mess was so traumatizing that the team outlawed use of the RV's bathroom. While there were bathrooms available in the mechanics shop, no other showers were available, so the guys went without washing for about six weeks. Then, in mid-February, one of their computer sponsors, Intel, invited the Nevada members of Red Team to San Francisco, where the computer chip manufacturer wanted to show off Sandstorm at the Intel Developer Forum.

By that time, Sandstorm had managed a speed of 49 mph and an autonomous run of a hundred miles. The guys were excited about the progress they'd made. But the robot still had its mechanical idiosyncrasies. It was apt to see obstacles that weren't there, or miss obstacles that were, or even misinterpret pre-programmed commands. What if something like that happened while Sandstorm was onstage at the conference?

The following morning, an audience of hundreds watched the autonomous vehicle creep out onto the stage, apparently thanks to the benefit of high-tech sensors, engineering and computers powered by "Intel inside." The crowd cheered in response. The applause felt good to the Red Team members present. Here they were at a Silicon Valley event being treated like celebrities. The recognition validated their sacrifices and the worth of the project. It also made the team thankful that no one realized that during the onstage demonstration, a Red Team member had hidden in the space under Sandstorm's steering wheel, prepared on a moment's notice to slam his hand on the brake pedal if the massive robot threatened to roll off the stage into the crowd.

On Friday, March 5, 2004—eight days before the race and just three days to go before the qualifying events—Chris Urmson rose early in the morning, put on his usual uniform of a mud-spattered baseball cap, fleece sweater and worn jeans, laced up his running shoes and decided that today would be the day to stage Sandstorm's culminating test challenge.

Urmson, Peterson, Spiker and the rest of the Nevada squad tested Sandstorm in the worst conditions they could imagine—frequently, along sections of the trail the old Pony Express had followed more than a hundred years earlier. "Red is really gung-ho about testing hard," explains Peterson. DARPA had said its route would be about 150 miles. The longest run Sandstorm had made was a hundred miles. But with the race a little more than a week away, everyone on the team was hoping for a longer run to boost their confidence.

The goal was just like the race: 150 miles in ten hours. The route amounted to a flat oval, about two miles around. While they prepared Sandstorm, Urmson and Peterson tinkered with a new part of the software: a component of the speed-setting module designed to slow down the robot when it approached a curve. The new code was designed to allow Sandstorm to drive more quickly on straight-aways.

The code worked wonderfully. During a few warm-up laps, Sandstorm managed to get up to 49 mph along the straightaways and then the new algorithm slowed it down as the robot headed into the curves. In fact, as Urmson and Peterson watched the robot, they wondered whether it slowed Sandstorm too much. An adjust-ment to the algorithm during a refueling break seemed to improve things. On the first lap they watched as Sandstorm cruised into a curve, slowed a little bit and then accelerated through the curve's exit. At the end of the *second* lap, Sandstorm was heading fast into what Urmson would later describe in his field test report as a "soft S-curve" to the left. The right-side tires drifted off the road into deep sand, and when Sandstorm tried to correct things, to get back on the track's packed-down dirt, it steered too hard to the left. The right-side tires bit into the soft sand. The left-side tires came up off the road. Behind, in the chase car, Urmson watched, horrified, as Sandstorm tipped up and over, and came to rest upside down—right on top of the e-box and the gimbal housing all the vehicle's most sensitive equipment.

The robot had been designed to insulate the box's components from being damaged in all sorts of accidents. Front-end collisions, rear-end collisions—pretty much any collision that happened on the ground plane, Sandstorm would be able to withstand just fine. But the robot had one fatal weakness: a rollover. Because Humvees sat comparatively low and flat, their geometry made rollover acci-dents almost impossible.

Unless you were testing a robot Humvee in the Mojave Desert, apparently.

A History Channel crew had come out to film the test run. They rushed out onto the track with their cameras and shoved one into Urmson's face, asking him to list the damage. Urmson looked at the wrecked robot the team had spent the better part of a year engineering: at the crushed gimbal, the compacted GPS antennae, the flattened e-box and the connecting rods bent out of shape. And he let fly with the expletives that made him one of the few people to ever have to have been bleeped by the History Channel. "Shock and disbelief," Urmson says when asked to describe his reaction, more than ten years later. "But mostly disbelief."

Disbelief, because they had felt like they'd been making such great progress. Disbelief because they were just days from the qualifiers. Disbelief because this had happened on the second lap of a two-mile track that Sandstorm was supposed to drive for seventy-three more laps.

Most of the crew figured Red Team was over. That they'd never be able to repair the robot in time. Somebody called Pittsburgh to inform everyone else about the accident. Red's assistant Michele Gittleman took the call. She recalls sobbing when she processed the news.

— — —

Maybe a crew led by someone other than Red Whittaker would have given up. But Whittaker didn't even consider it.

At the Nevada Automotive Test Center, Urmson, Peterson, Spiker and the rest of the team attached the four-by-four chase vehicle to Sandstorm with the help of some nylon webbing and managed to flip over the robot. Spiker, the most mechanically minded among them, went over the engine to look for problems. The engine was flooded with diesel fuel but aside from that, everything looked okay.

The other guys assessed the electronics equipment. The GPS units were toast. The gimbal suffered the worst impact and would need to be completely rebuilt. The main LIDAR unit was irreparable. Luckily, an extra gimbal and LIDAR sat in storage back in Pittsburgh. They towed Sandstorm to the mechanics shop and for three nearly sleepless days and nights they worked to fix everything they could. And they nearly did it.

The race was March 13, 2004. Heading into the qualifiers the week before, at the California speedway in Fontana, the GPS system worked, which meant the robot could locate itself in the world. The sensors were active, which enabled Sandstorm to perceive obstacles. The computers could calculate the trajectories required to follow the path set out by the Red Team mapping crew. The only problem? "We had no time to calibrate," recalls Whittaker. Which meant Sandstorm viewed the world through a distorted lens.

Think of each sensor as its own individual eyeball. You are able to see one version of reality because your brain is able to amalgamate the view from your two eyes into a single picture of the world. Sandstorm amalgamated the information from four different LIDAR units plus the stereo-camera system. Ensuring that the robot's sense of the world resembled the actual world required calibrating the individual sensors—a time-intensive process of trial and error. "Think of calibration as alignment," Whittaker explains. "Even in the car shops they align your headlights, right? And when there are multiple sensors that will fuse data into a common model, it's important that they're all aligned. If you just bolt it back together you're creating a kind of Frankenstein, and maybe it's a little cross-eyed."

And so the cross-eyed Frankenstein's monster limped into race week, belching diesel exhaust, dented and scratched, but otherwise intact. Red Team would compete against twenty other entries from across the United States. In the qualifiers each of the twenty-one robots would have to navigate a mile-long obstacle course to progress to the main event.

Soon after his arrival at the California Speedway, Urmson wandered around to see what he could learn about his competition. He saw Doom Buggy, created by the only high school admitted to the competition, Palos Verdes, near Los Angeles. An undergraduate from UC Berkeley, Anthony Levandowski, led the team behind the only two-wheeled entry, a robot motorcycle that was able to balance itself with the help of a gyroscope. UCLA's entry, the Golem Group, was led by a guy named Richard Mason, who had seeded his project with $28,000 he'd won on *Jeopardy!*

On the other end of the spectrum were the professional teams, which were affiliated with various engineering-focused corporations. An inventor named Dave Hall had created an autonomous

Toyota Tundra pickup truck that was notable for driving smoothly with a stereoscopic camera setup—it didn't use any LIDAR at all. From Wisconsin, the makers of Oshkosh Trucks entered a six-wheel-drive, 32,000-pound fluorescent yellow behemoth with the imposing name of TerraMax. Louisiana's Team CajunBot also used a six-wheeled vehicle. A fraction the size of the Oshkosh entry, it was based on an all-terrain vehicle more commonly used by the state's hunters to navigate the bayou.

Urmson, Peterson, Whittaker—all of them wandered the event, talking to people just as technically minded as they were. It quickly became apparent that Red Team was among the biggest of the teams. *Popular Mechanics* gave them seven-to-one odds to win, highest of all their competitors'. The front-runner status positioned Red Team as the entry everyone else wanted to beat. Urmson had posted on the Red Team website a photo of Sandstorm after the rollover. Now, as he wandered the raceway, talking to the leaders of other teams, Urmson spied the photo on numerous computer monitors. Some other teams had made it their wallpaper—as motivation.

Adding to the excitement was the fact that DARPA's public relations team had arranged for reporters and television producers from across the nation to visit the raceway. Urmson and his teammates had toiled for months in obscurity. The accolades they received at the Intel event had been nice, but the more common reaction to their work was incredulity. "A car that drives itself?" people would scoff. To many, it sounded ridiculous. The presence of the reporters going around interviewing anyone available reminded the competitors that their work was important. Important enough for the U.S. government to put up a million-dollar prize. Important enough, possibly, that it might save the lives of U.S. soldiers fighting in distant theaters of war.

On the morning of March 13, 2004, the start of the race was one of the most exciting moments Chris Urmson had ever experienced. The robots were lined up in their starting chutes. Media and military helicopters hovered in the sky. Grandstands supported hundreds of spectators, each of them getting whipped by the desert sand, and over it all, Tony Tether's amplified voice marked the momentous event.

"We're thirty seconds from history," shouted the DARPA director into the microphone. "All right, ladies and gentlemen, boys and girls, the bot has been ordered to run, the green flag waves, the strobelight is on, the command from the tower is to *move!*"

Because Sandstorm had performed best in the qualifiers, it had the honor of starting first. The big Humvee rolled slowly out of its chute. "Ladies and gentlemen, Sandstorm!" Tether cried. "[An] autonomous vehicle traversing the desert with the goal of keeping our young military personnel out of harm's way."

The first complication in the race course was a leftward turn. Its inside edge was marked with some scrubby vegetation, and its outside edge, with concrete jersey barriers protecting spectators from the robots. Sandstorm followed the road perfectly throughout the curve and accelerated once it headed out on its straightaway.

While it was still in view, Sandstorm ran over a hay bale. Urmson winced. But the big off-road vehicle just kept on going. Soon, the Red Team couldn't see their robot at all. No one had thought to provide the teams with a video feed of their vehicle's progress. All they could do was settle in and wait to hear reports issued back to the start from helicopter-borne observers and other officials set along the course.

Soon, the other entries headed out: A team called SciAutonics II. Then CajunBot rolled its six wheels from the starting chute and drove straight into a jersey barrier. Team ENSCO's robot, based on a Honda ATV, wandered from the road just past the turn, flipped over on its side and was out of the race just two hundred yards into the event.

Palos Verdes High School's autonomous SUV also ran into a jersey barrier. And then came the most curious of the entries: Anthony Levandowski's autonomous motorcycle. Levandowski pulled it up to the starting line, activated the gas-powered motor, stepped away—and watched, brokenhearted, as the motorcycle immediately tipped over. As Levandowski would discover later, he'd forgotten to activate the gyroscope that kept the motorbike balanced. His race was over.

Minutes later, Red Team heard from a race organizer that something was wrong with Sandstorm. The hay bale the robot had run over just after the start turned out to reflect an ongoing problem. Perhaps because its sensors hadn't been calibrated properly, perhaps because the main LIDAR's replacement unit scanned at a much slower rate than the original, Sandstorm consistently appeared to think that it was a foot or two to the left or right of where it actually was. The Humvee drove over a fencepost, then another and a third. Some miles later, the vehicle swung itself into a curve, a particularly tricky one given the inside edge was separated from a steep drop-off by only a knee-high berm. As Urmson and Peterson intended, Sandstorm slowed down as the road turned. But the robot was a foot or two to the left from where it should have been. As a result, the left-most tires climbed up the berm, then dropped down the steep inside ledge. Sandstorm was now stuck on its belly—what Urmson called "high-centering."

Things quickly grew worse. Sensing Sandstorm wasn't moving, the speed control system kicked in, directing more power to the engine. One of the tires hanging over the other side of the berm was situated just high enough off the ground that it could still touch the Mojave sand. The friction heated up the rubber until it smoked and eventually burst into flames. The robot's progress was over 7.3 miles from its start.

The media used Sandstorm's flame-out as a metaphor for the entire event. The number-two entrant, SciAutonics II, also got stuck on a low hill of earth. Dave Hall's Toyota Tundra became confused by a small rock. The UCLA entry, Golem Group, stalled out when a safety device prevented its engine from accelerating enough to get up an incline. And TerraMax, the 32,000-pound monster truck known for its brute force approach, ended up halted when a pair of tumbleweeds it incorrectly considered immovable obstacles blew ahead and behind it. And those were the best-performing vehicles.

The result put DARPA director Tony Tether in a tough spot. At the other end of the race course, in Primm, Nevada, was a tent full of reporters who had traveled across the country to file stories on the race winner. Tether figured he was going to get killed by the press—an expectation that proved right. "DARPA's Debacle in the Desert," went one headline. The gist of the stories portrayed DARPA as an out-of-touch government bureaucracy that had wasted money staging a fool's errand. So to distract them, Tether took the stage and announced a second race, to be held in a year or so, with a doubling of the 2004 race's purse, to $2 million.

Chapter Two

A SECOND CHANCE

*The only way to prove you're a good sport
is to lose.*

——ERNIE BANKS

Red Whittaker started planning for the second race even before
Sandstorm returned to Pittsburgh from the first. Through his
repeated entreaties for sponsorship, Whittaker had developed a relationship with AM General, the company that manufactured the
Humvee. Now Whittaker thought he could convince the executives
to donate an additional vehicle for Red Team to use in the next
challenge—if the executive team would only witness a demonstration of Sandstorm's capabilities.

Several days after the first challenge, Whittaker, Spiker and Peterson arrived with Sandstorm at the AM General campus in South
Bend, Indiana, to conduct that demonstration. Spiker and Peterson
stayed outside and set up the robot on an obstacle course the Humvee manufacturer maintained to educate new owners on the capabilities of their vehicles.

One element of the obstacle course was a concrete tabletop structure, maybe eighteen inches off the ground. Peterson and Spiker
wondered whether Sandstorm could drive itself up and onto the
obstacle. Moments later, rather than creeping toward the tabletop,
as Spiker and Peterson had intended, Sandstorm took off toward
it at high speed.

A kill switch was designed to deactivate Sandstorm if it ever did anything unpredictable. Trouble was, the kill switch had about a two-second delay. Spiker pressed the switch, but Sandstorm hit the tabletop before the command took effect. The front wheels bounced the front end into the air. The rear wheels hit the tabletop and bounced up the Humvee's back end. For a moment the entire vehicle was airborne. Then the front end nose-dived with a violent slam against the concrete.

That's when the kill switch disabled the vehicle.

Spiker and Peterson rushed to assess the damage. Whittaker was in a nearby building conducting his presentation for AM General executives on Red Team, and the wonderful capabilities of the robot they'd developed. Outside, Spiker and Peterson discovered the impact of Sandstorm on the tabletop had crushed an engine-compartment coolant tank. Once that was repaired, they set up Sandstorm on a section of clear road and activated the giant robot to test it. Immediately the front wheels turned to the right. That shouldn't have happened. "Kill kill kill!" Spiker shouted to Peterson. With a snort of exhaust, Sandstorm accelerated right off the road and straight into the building where Whittaker was talking to the AM General executives. The impact of the Humvee against the wall shook the entire structure.

Later, Spiker figured out that the tabletop collision had detached a steering position sensor from its mooring—which, in turn, caused the second accident. But it turned out not to have mattered. Whittaker and the AM General executives rushed from the building to investigate the source of the impact. Spiker figured the sponsorship bid was toast. But as the execs surveyed the scene of the accident, Spiker realized his fears were groundless.

"Unflinching grace" is the way Whittaker characterizes the AM General execs' reactions, portraying them as "great hosts who don't fuss over a dropped fork or spilled water." The executives saw themselves as manufacturing a vehicle designed to push the bounds of what an automobile could do—and so, in its own way, did the Red Team. Of *course* they would sponsor Whittaker's team. "We'll give you *two* Humvees," one of the AM General execs proclaimed. "Just be careful."

— — —

Some months later, in the summer of 2004, a computer scientist named Sebastian Thrun listened to a presentation about the first DARPA Grand Challenge in a seminar room at Stanford University. Thrun had recently moved from a faculty position at Carnegie Mellon's Robotics Institute, where he'd been working on a project with Red Whittaker—a robot called Groundhog that was designed to map Pennsylvania's abandoned coal mines. His new job was in Palo Alto, California, leading the Stanford Artificial Intelligence Laboratory, a once-respected research facility established by AI pioneer John McCarthy in 1963, which had been dormant since it had been rolled into the greater computer science faculty in 1980. To reincarnate the facility, Thrun brought nine Carnegie Mellon academics with him. Having left behind all his projects at his old school, Thrun was looking for a quick way to reestablish the AI lab's reputation.

Thrun had attended the first Grand Challenge as a spectator, and was intrigued by the prospect of entering the second, as the rebooted Stanford AI lab's first major feat. So Thrun asked one of his fellow CMU transplants, who had also attended the first challenge, to conduct a presentation to the rest of the group.

The presenter was Mike Montemerlo, a soft-spoken engineer who had a reputation as a software whiz known for his ability to program robots to conduct the simultaneous localization and mapping that had so bedeviled Sandstorm in the first race. Montemerlo's father, Melvin Montemerlo, was a program executive at NASA and had worked closely with Whittaker on numerous projects. When Mike had been in high school, his dad had taken him on a pre-college trip to experience firsthand candidate campuses. One evening in Pittsburgh, the pair of them threw pebbles up at Whittaker's window to convince the robotics legend to give the teenager a tour of the Field Robotics Center. That experience was the reason

Montemerlo attended CMU. Years later, Whittaker would become Montemerlo's PhD adviser; in the same period, Montemerlo also happened to be Chris Urmson's officemate.

At Stanford, Montemerlo's presentation amounted to a travel-ogue of his experiences at the California Speedway. Full of photos of the various robots, the seminar highlighted the problems and foibles that each team experienced. He spent a lot of time on the work that had almost been destroyed by Sandstorm's rollover ac-cident. The penultimate slide asked whether the Stanford AI lab should compete in the second DARPA Grand Challenge. The final slide featured the answer: "No," in bold and all caps.

Thrun is a slim man who communicates in perfectly enunciated, precisely formed sentences colored with a German accent; he was born in the small Rhineland city of Solingen and raised in north Germany. "Why not?" he asked softly.

"It's hard," said Montemerlo, whose side-parted brown hair and wire-framed circular glasses made him resemble the Hollywood ste-reotype of a software engineer. "It's all encompassing," he followed up, perhaps thinking of the experience of Urmson and the rest of the CMU team. "People have to work all day and all night. They lose their social life. And—it can't be done!"

Somewhere, somehow, Montemerlo must have known that tell-ing Thrun that something couldn't be done was the quickest way to entice him to try it. "I'm a rule breaker," Thrun says, a character trait he shares with Whittaker. "A rebel—I like to do crazy things."

Thrun was the third of three children. "I was the one the parents didn't have the energy and time to pay attention to," he told one re-porter, years later. "I remember a beautiful childhood—but pretty much on my own." Left to his own devices, he developed various ob-sessions with intellectual projects. At the age of twelve, in 1980, the obsession involved a Texas Instruments pocket calculator that could be programmed to solve various equations. Thrun delighted him-self using it to create little video games. Next, he happened upon a Commodore 64 personal computer on display in a local department store. The computer was too expensive for his middle-class family, so Thrun returned to the store display to program on it, day after day, week after week. Each day he tried bigger and bigger programming challenges. He grew adept at efficient coding; because the staff turned off the computer each night, he had to execute each challenge he set himself in the two and a half hours that passed between the end of the school day and the store's closing time.

By the time Thrun's parents bought him a used NorthStar Horizon personal computer, the young man was able to program simple video games. He wrote a virtual simulation of the Rubik's Cube. Another feat involved coding the member database for his family's tennis club. One gets a sense that Thrun roved through his adolescence seeking out challenging problems that he would use to test his programming ability. The same method would predominate in Thrun's academic and professional life. He enrolled in the computer science department at the University of Bonn. Artificial intelligence attracted him because, in comparison to humans, with their sometimes irrational, inscrutable behavior, Thrun felt he could fully grasp the reasons a software program acted the way it did.

In 1990, the University of Bonn bought a Japanese robotic arm as a research tool. Thrun distinguished himself by using a neural network to teach the robot how to catch a rolling ball. The resultant academic paper was accepted to an American artificial intelligence conference, Neural Information Processing Systems. The trip was a turning point for Thrun, who was then twenty-two. He'd discovered people exactly like him—a whole community of "psychologists and statisticians and computer scientists all working together to understand how to make machines learn." From that moment on Thrun focused on writing academic papers so he could attend more AI conferences. Through such gatherings, Carnegie Mellon AI legend Alex Waibel became a mentor, as did Thrun's future thesis adviser, Tom Mitchell. Thrun joined the CMU faculty after he earned his PhD in computer science and statistics from the University of Bonn in 1995.

One of the most interesting projects Thrun worked on in Pittsburgh was the creation of a robot tour guide for museums. In keeping with the kitsch factor the public associated with robots—think the 1986 comedy *Short Circuit*, the TV show *Knight Rider* and Data, the well-meaning android on *Star Trek: The Next Generation*—the tour guide that Thrun constructed, Minerva, included a pair of camera lenses for eyes and a red mouth that could tilt into a frown to indicate displeasure. As a publicity stunt to demonstrate the capabilities of technology, Minerva even provided tours to visitors of Washington's Smithsonian Museum.

It turned out programming a robot to navigate through a museum was a surprisingly complex challenge. Minerva would share the museum floor with dozens of human tourists—as well as valuable museum exhibits. How to engineer the creature so that it didn't bump into an exhibit? How to write the code so that it didn't roll over a child?

Six years before DARPA staged its first Grand Challenge, in 1998, Thrun equipped Minerva with laser-range finders. Then he loaded the robot with a machine-learning algorithm and sent it out on the museum floor at night, without any tourists around. Minerva wandered around the exhibits, sending out laser beams and creating a map of its environment. Then, when the museum was open, with humans sharing the same floor as the robot, Minerva would use this map to locate itself. The map also provided a way for Minerva to avoid running into humans. The robot would assume any new obstacle that hadn't been on the original map was a human, causing Minerva to stop safely.

The tour guide was a big hit, and Thrun used the acclaim to handle the software side of other projects. For example, Whittaker convinced Thrun to join the team that built the Groundhog robot that aimed to make it safer for Appalachian coal miners to retrieve their underground ore. Maps didn't exist for older, decommissioned mines in the area, which could cause problems. In 2002, for example, nine workers toiling in Pennsylvania's Quecreek mine were trapped by water when they breached an adjacent passageway that had been abandoned for years and flooded sometime along the way. The miners escaped after three days, but Whittaker took the accident as a challenge and, in just two months, with Thrun working on the SLAM programming, created a robot that could be dropped into old mines to scan the passageways and create 3-D maps for reference.

DARPA's series of challenges fascinated Thrun. When Thrun was eighteen, in 1986, his best friend, Harald, was invited for a ride in another friend's new Audi Quattro. It was an icy day, and the driver was going too fast and ran the Quattro headfirst into a truck. Harald died instantly. The impact was so strong that his seat belt was shredded. The crash would forever haunt the German robotics professor.

Thrun saw self-driving cars as a way to make automobile transportation safer, to avoid crashes like the one that killed his friend. He did some thinking about the problem after the first Grand Challenge. The fact that DARPA created waypoints along the route really simplified the problem, he figured. Programming Minerva to navigate the fast-changing and crowded environment of the Smithsonian Museum rivaled the complexity of the self driving-car problem. Before he left Carnegie Mellon, he went to Red Whittaker with an offer. "Look," Thrun told the older robotics legend. "I've been recruited from Stanford, but for the next Grand Challenge, I would love to help you."

"Had he said yes," Thrun recalls, "I would have happily served on his team and never have started my own team."

But Whittaker declined Thrun's offer, presumably because he wanted to keep Red Team exclusive to people associated with Carnegie Mellon. After Montemerlo's presentation, Thrun considered whether to enter the second challenge himself. Red Team had taken a year to build a robot that went 7.3 miles. If Thrun's new lab could do better, they'd go a long way toward establishing a national reputation. SLAM would be integral to a successful performance, and Thrun and Montemerlo were two of the world's leading experts on the topic. Thrun basically figured, why not?

So on August 14, when DARPA staged a conference for potential competitors, Thrun brought Montemerlo and several other members of his team. The conference was held in Anaheim, California. Despite the negative media coverage of the first race, even more competitors came out this time around: more than 500 people from 42 states and 7 different countries attended the 2004 competitors' conference. Ultimately, 195 teams would register to compete, nearly double the number that signed up for the first race.

Including, of course, the Red Team. The summer after the debacle in the desert, Urmson went off and completed his PhD, then got a job working for Science Applications International Corporation, the government contractor that had sponsored Sandstorm. Urmson's assignment was to work with Red Whittaker and Red Team on the second DARPA race. Urmson's hopes were considerably higher for the second challenge. They'd have another eighteen months to perfect Sandstorm's development. And they'd be doing so with a more professional group, including several engineers from Caterpillar, the construction-equipment manufacturer. The budget was bigger, at $3 million. The atmosphere was different, too. The first time out there was youthful enthusiasm. This time, there was an almost grim determination.

"I signed up to win the Grand Challenge," Whittaker proclaimed. "This time around, the Red Team will be more like a Red Army."

It was inevitable that the Stanford and Carnegie Mellon teams would bump into each other at the preliminary conference. Urmson noticed that Montemerlo was carrying a sheaf of papers in his hand that turned out to be the technical paper Urmson had written after the first race. The paper described the most intimate details of Red Team's approach. Publishing for the rest of the robotics community the secrets of all competitors' approaches had been one of DARPA's conditions of entry. It was a good strategy. In the spirit of academia, sharing intelligence meant the whole field progressed faster. But it also made things more difficult for Whittaker and Urmson. As the country's leading robotics lab they'd had a head start for the first race. Publishing their approach brought everyone else closer to the Red Team's level. And the defectors, Montemerlo and Thrun, were brilliant people. That they were entering meant the prize was no longer Carnegie Mellon's to take. Now, heading into the second challenge, Red Team faced its most serious competition yet.

— — —

Early on in its preparations, Red Team decided to hedge its bets by entering two robots. (There was a precedent for this. SciAutonics had entered two vehicles in the first race.) Partially, the step was designed to smooth relations between team software lead Kevin Peterson and project manager Chris Urmson, who were apt to butt heads in the latter half of Sandstorm's development. There was talk of giving each deputy his own vehicle, although years later Whittaker would insist that Peterson and Urmson contributed to both robots in the lead-up to the second race. And partially, the move was pragmatic. After all, thanks to AM General's donation, Red Team had enough Humvees.

The second vehicle, which became known as H1ghlander, was a 1999 model year, making it thirteen years younger than Sandstorm. The AM General–donated vehicle came with a 6.5-liter turbocharged engine. One of the challenges of autonomous driving involved controlling acceleration and steering. Most vehicles of the era were mechanically controlled. They relied on a human being twisting steering wheels, pushing accelerators, shifting gears, which complicated matters when a computer was supposed to do the driving. There was a margin of error when a digitally controlled actuator pressed against, say, a gas pedal.

This new Humvee, H1ghlander, featured drive-by-wire capability embedded in its controls. It had been designed to be controlled by a computer. The throttle, for example, was operated by a factory-installed engine control module. So instead of rigging up an electric motor and lever to actually push against the gas pedal, as with Sandstorm, the H1ghlander crew could hack into the newer Humvee's computer system and control the throttle electronically. It all meant less margin of error, which made H1ghlander a better driver.

Another change was that Whittaker and his students had tracked down a different, more accurate location-tracking system. The system used in the first race had a margin of error of about a yard. This new one, from a sponsor named Applanix, featured a margin of error of about twenty-five centimeters, or less than a foot—a big improvement for the second race.

So the Red Team had a lot going for it. But so, too, did Thrun's team. In his heart, Whittaker was a hardware guy, who came from an era when making robots work involved the precise interplay between actuators and carburetors, electric motors and solar-powered chargers. This was reflected in Red Team's approach to the first challenge, which saw his charges spending as much time perfecting the e-box and gimbal mechanisms as writing code for the computers. But as computing power improved, robotics was increasingly becoming a software problem, which computer scientists, rather than mechanical engineers, had to solve. Whittaker was an engineer. Thrun's team was dominated by computer scientists. Very little of the hardware that Stanford used needed to be custom-designed. In contrast to Sandstorm's gimbal and e-box, which the Carnegie Mellon team had engineered itself, Thrun simply took sensors he found in the marketplace and bolted them to his team's vehicle, including five LIDAR units, a color camera to aid road detection and two radar sensors designed to identify large obstacles at long distances. The philosophy of the Stanford team was to "treat autonomous navigation as a software problem."

"My perspective was, you take a human out of a car, and replace it with a robot—there's a bit of a hardware issue," Thrun observes. "You have to figure out how to crank the steering wheel and press the brake. But that part is trivial. You put a little motor on the steering wheel. There's no science . . . It's all about artificial intelligence. About making the right decision. So we had this complete focus on making the system *smart.*"

"Carnegie Mellon was a team—it's a humongous place, and they have experts in everything," Montemerlo explains. "We were a much smaller group. We very much were software people. None of us had any mechanical skill whatsoever."

That said, Thrun had learned a lot from his experiences working for Whittaker. In September of 2004, fresh off the heels of Montemerlo's presentation, Thrun used Whittaker's template to begin work on his own entry in the second DARPA Grand Challenge. Just as Whittaker did, Thrun recruited volunteers by asking them to enroll in a university class. Thrun's was called "Projects in Artificial Intelligence." At the first meeting of maybe forty students Thrun gave a Red Whittaker–style inspirational speech. "Look, there's no syllabus, no course outline, no lectures," Thrun recalls saying. "All we're going to do is build a robot. A robot car that can drive on the original course."

Thinking of the way Whittaker motivated his students to work hard by providing them with challenges, Thrun set his class a clear and well-defined objective: By the end of the two-month-long session, they were to have built a car that could travel a single mile of the first DARPA Grand Challenge course. "Red and I have very different personalities," Thrun says. "But I tried to learn from him. And what I learned from Red was, when you give students a goal, no matter how hard it is, because they haven't learned that these goals are hard to reach, these students think they can reach it. And eventually, they do reach the goal."

The class didn't have a budget to go out and buy a car. Someone contacted Ford to ask the manufacturer to donate one, and the company said yes, but they wanted it back afterward, in the same condition they lent it out. Perhaps thinking of Urmson's rollover accident, Thrun declined the Ford offer. Luckily, a friend of his named Joseph O'Sullivan, an AI researcher who worked for Google, played soccer with a guy, Cedric Dupont, who worked as an engineer at Volkswagen's lab in Palo Alto. Dupont arranged to provide Thrun's team with a 2004 Touareg R5 TDI, as well as the help of VW engineers to access its computer system. "That was like a gift from God," Thrun says. Like Highlander, the Touareg had a drive-by-wire interface, and with VW's help, Thrun's team could hack into the computer system relatively easily.

Thrun ended up with about twenty people committed to joining the Stanford team, which he split into smaller units. One group was charged with configuring hardware—actually attaching the sensors to the Toureg, which, in a nod to their school, they gave the nickname Stanley. Another part of the team was in charge of providing the mapping. A third handled navigation.

Two months later, at the end of the term, Thrun took his students out to the Mojave Desert and set up Stanley on the course of the first Grand Challenge. Then they activated the robot and watched: Stanley drove past the class's one-mile goal, thrilling Thrun, who became even more excited when Stanley passed 7.3 miles, which was how far Carnegie Mellon's Sandstorm had made it. Some minutes later, at 8.4 miles, Stanley found itself stuck in a deep rut, caused by heavy rain.

Thrun was beside himself. The sort of rut that had stymied his robot would have been smoothed over by DARPA prior to the race. Had this been an official race day, it's possible Stanley would have proceeded much farther. "That was just unbelievable," Thrun recalls. "That was the moment it became clear to me, boy, there's a real possibility it can be done." If a team of comparative novices could surpass the best Carnegie Mellon team in just two months, Thrun wondered, then what could the same team do in the year leading up to the second race?

—

Red Team's strategy this time around amounted to a bigger and better version of the approach they'd intended to execute in the first race.

Truth be told, they felt a little cheated by the way the first race went. The communication out of DARPA had led the team to believe that the robots would have to navigate rough territory and brutal off-road conditions. DARPA's actual route turned out to have some hairy spots, such as tunnels and narrow fence gates. But there was nothing arduous about the road itself. That had been a smoothly graded desert thoroughfare. Your typical subcompact import could have driven off a car lot and navigated it. Looking back, Red Team had wasted countless hours ensuring their robot would be able to handle off-road conditions. And not just handle them—handle them *fast*. That's why they'd used shocks and springs to suspend the electronics box and the gimbal, to ensure the computer equipment would be able to withstand the resulting jars and vibrations. Had Red Team forgotten about testing the robot in the most difficult of conditions, and just concentrated on developing a vehicle that would be able to roll from one GPS waypoint to another, then many team members figured they would have ended up finishing the first race. They could have won.

So this time, Whittaker concentrated on refining the capabilities Red Team had already developed, including the pre-driving approach that it had used in the first challenge. In August 2005, Whittaker moved both Sandstorm and H1ghlander out to Nevada. The robotics engineer figured the federal agency would amp up the difficulty for the second race. Some of the toughest roads in the nation were the M1 Abrams tank courses at the Nevada Automotive Test Center. So that's where Red Team landed with just three months to go, to put the robots, and the team, through a series of what were in effect dress rehearsals designed to replicate race-day conditions—right down to special costumes worn by DARPA staff stand-ins.

Red Team tended to use two different routes to test its vehicles. One, known as the "Pork Chop," was a 48-kilometer loop that featured everything from dirt road and pavement to cattle guards, high-voltage power lines and railroad crossings. The Hooten Wells route was an 85-kilometer one-way line that followed the course of the Pony Express and featured a dry lake bed, gravel road and a narrow canyon.

The testing featured its share of disasters. Spiker had a credit card linked to a Carnegie Mellon account and was authorized to spend $100,000 a month, a figure he regularly blew past procuring the spare parts required to repair Sandstorm and H1ghlander after the damage caused on their testing runs. For example, on August 26, just twelve days after they arrived in Nevada, H1ghlander sheered off its front right wheel as it navigated a particularly treacherous off-road trail. On September 15, Sandstorm was clotheslined by a tree, sustaining significant, but nevertheless repairable, damage.

These setbacks aside, the testing was going well.

For the first time, Sandstorm and H1ghlander were completing challenge-length runs that featured some of the toughest terrain the team could throw at the robots. The vehicles drove more than 1,600 kilometers each. Better yet, they were completing these runs in times that would have them finishing the race in under seven hours. Red Team was feeling very good about its chances.

Even so, Whittaker was working his team as hard as he ever had. The 4:00 A.M. wake-ups were taking their toll. The race rehearsals started at 6:30 A.M., just like they would during the actual event, and then, after the course work, the team would take the robots back to their garages, where the coders and the mechanics would work long into the night to make improvements and repairs. The next day, they'd rise at four and do it all over again. It was a grueling routine. "Everyone was scraped raw by exhaustion," Whittaker recalled.

To refresh everyone, to ensure his team was sharp and fully rested come race day, Red set a week's vacation before the national qualifying event, which began September 28, 2005, at the California Speedway. There, forty-three teams would be evaluated by DARPA, competing to become one of twenty-three finalists to qualify for the actual race on October 8, 2005.

The final day of testing was September 19. Whittaker's culminating goal had Sandstorm and H1ghlander navigating 10 laps of a 30-mile-long course, to accumulate 300 miles in total, about double what the robots would have to do on race day. Once they achieved the distance, the team would freeze the software, store the robots and disperse to their own chosen habitats for the pre-race rest.

By the afternoon of the nineteenth, Sandstorm was ready for the race but for a last-minute tire and oil change. Meanwhile, H1ghlander was nearing the final laps of its last test session. Following behind in AM General's second donated Humvee was Peterson in the passenger seat and software engineer Jason Ziglar behind the wheel. Ziglar was doing his darnedest to keep up with H1ghlander, whipping the steering wheel this way and that, his foot jammed on the accelerator. With H1ghlander about to start its final lap, having already gone 270 miles, Peterson called Red in Pittsburgh, where he was handling some last-minute details. "The vehicle is driving really well," Peterson told him. "But we're really beating up on it." What if something happened? Peterson recommended to Red that they call off the final lap. "It felt like we'd learned everything we were going to learn," Peterson recalls.

Giving up before the team had completed a goal wasn't in Whittaker's DNA. He made the call—*finish the route*. So they kept going. Moments later, H1ghlander was kicking up its usual dust cloud. From the passenger seat in the chase vehicle, Peterson couldn't see the robot, but thanks to his laptop's Wi-Fi connection, he could see what H1ghlander could see on the monitor. Approaching a leftward curve, the robot slowed down, the way its algorithms specified, and then accelerated into the curve. Except it swayed just a little bit to the right, off the path—and Peterson's whole display went red. When the dust cleared, Peterson saw a dirt formation on the right side of the road that looked like the sort of thing a stunt driver would use to shoot a car up into a two-wheeled drive. In this case, the stunt jump had sent H1ghlander over on its side, and ultimately, onto its roof. The robot had caught the right wheels on the ramp at 30 mph and launched itself into the air.

Another rollover.

Having been through this before, the team leapt into action. No one broke down in tears over this one—Spiker was prepared. Many of the extra parts required to repair H1ghlander lay in the mechanics shed at the Nevada Automotive Test Center base. The rest, Spiker arranged to have shipped from Pittsburgh to Nevada.

And that week's worth of vacation everybody was supposed to go on the next day? Gone. Instead it turned into the biggest work session the Red Team had ever faced.

Once Stanford's AI class conducted its 8.4-mile test run, Thrun winnowed his team down to four key people. Thrun himself and Carnegie Mellon alum Mike Montemerlo were the first two. Among those who had taken his robot class, Thrun discovered a fellow German, a computer-vision expert and programming whiz named Hendrik Dahlkamp. The fourth was a grad student named David Stavens.

A quartet was appropriate to the task because that's how many occupants the Touareg comfortably fit. For a week at a time, Thrun and the other three would head out into the Mojave Desert and drive the trails. At first, they'd set the vehicle on a trail, watch it navigate itself, and eventually the robot would encounter something it couldn't handle. Then someone would code a fix. As the process repeated itself dozens, and eventually hundreds, of times, the robot became sophisticated enough that it began to teach itself. In this phase, Thrun would drive Stanley through the desert, manning the controls, slowing down when the road became rough or steep, accelerating on smooth straightaways. After several days of this, Thrun would go back to the university, and Stanley, working overnight, would retroactively look at the data to engage in its own learning. Confronted with *this* terrain, Stanley would think, Sebastian chose to drive *here*—and I will do the same. "The robot would basically spend the night sorting through the data and bring order from chaos," Thrun said.

Stavens's contribution was an algorithm that taught the robot how to regulate its speed. The roads Stanley drove in the Mojave featured rain ruts, puddles and potholes. Blasting through this sort of terrain at speed would have shaken the car to pieces. So Stavens wrote a program that regulated Stanley's progress based on vibrations felt by the robot's sensors, as well as the grade and width of the road. With the program loaded into the robot, Mike Montemerlo drove Stanley to create data the program could then analyze to develop rules that would guide its behavior.

The problem here was that Montemerlo was too conservative. He's incredibly detail oriented. A nice way of putting it is risk-averse. "We used to put stickers on his windows," Thrun recalls. "So

Mike couldn't see how fast we were going." Montemerlo had once protested to the team members that he would never get in a self-driving car that went more than 5 mph. Driving Stanley, Montemerlo would creep around the desert, easing up hills, wandering over rubble and stones. Then, once the vehicle was at home, the machine learning algorithm would look at the way Montemerlo drove and create rules that would guide Stanley in the future. Accustomed to high-speed driving on Germany's Autobahn, Thrun didn't like how slowly Stanley progressed once it had crunched Montemerlo's data. So one week, when Montemerlo went away on vacation, Thrun set Stanley to go 20 percent faster.

Then came the day in 2005 when Thrun received an unexpected visitor at his Stanford office. He looked up and saw a figure in the doorway. The figure came forward and introduced himself: "Hi," the man said. "I'm Larry Page."

Thrun knew who Page was, of course. What surprised him was how interested Page was in the project. "Larry's always been a robotics enthusiast," Thrun says, explaining that had Page not started Google, he might have pursued a PhD in robotics. Page was fascinated with Thrun's project. He had about a million questions. He wanted to see how real the technology was—how close are driverless cars? A century? Decades? A couple of years? What did Thrun think? In fact, Page was so interested that he told Thrun he planned to attend the second Grand Challenge. Through their shared enthusiasm for driverless cars, Thrun and Page developed a friendship that deepened, because the two men both relished taking on tasks that everyone else dismissed as impossible. Thrun had no idea, at that point, that Page would change the course of his life.

At 4:30 A.M. on October 8, 2005, the day of the race, DARPA officials provided a Red Team member a USB key featuring a computer file of 2,935 waypoints—the course of the second Grand Challenge. The whole of the route totaled 132 miles, starting and ending in Primm, Nevada.

The next bit bore many similarities to the first race. The team member sprinted to Red Team's command center. Another member loaded the route network definition file onto Red Team's shared hard drive. A computer program analyzed the waypoints and added thousands more, so a route originally specified every eighty yards now featured a dot every yard or two. Next, the route was divided up among team members to go over. The pre-planning team went through each part of the route to ensure the new waypoints kept Sandstorm and H1ghlander on navigable road.

In the anxious moments that passed while the pre-planning team worked, Whittaker, Urmson and Peterson discussed strategy. The experience of the first race eighteen months before was fresh in everyone's minds. That time, they'd gone for speed. And perhaps they'd pushed Sandstorm beyond what was good for it.

So the three decided Red Team should take a tortoise-and-hare approach with its two vehicles. One of the vehicles would take it easy, going so slowly that it would be *certain* to finish the race. This way, in the event that no one else finished, at least Red Team would have a vehicle that crossed the finish line.

Sandstorm consistently came in 10 percent slower than H1ghlander—a symptom, the engineers thought, of the way the electronics box floated, which made it difficult for the robot to pinpoint exactly where it was. So H1ghlander would be Red Team's hare, while Sandstorm was the tortoise.

In terrain the pre-planning team classified as *moderately difficult*, H1ghlander would go 20 percent faster than Sandstorm. In *very safe* territory, Whittaker decided that Sandstorm was allowed to go 27 mph, while H1ghlander was able to go up to 30 mph—an increase in speed of 12.5 percent. H1ghlander, Red said, should target to finish in 6 hours and 19 minutes, for an average speed of 21 mph. And their safety, Sandstorm, should finish in 7 hours, 1 minute.

Urmson and other Red Team members watched the race from Stanford's tent, because Stanford had the best view. H1ghlander was first out of the starting chute. And in the initial few miles, it led the pack. Then, nearly seventeen miles in, H1ghlander faltered. The engine stalled and the vehicle came to a stop, then started again. Coming up on a hill, it stalled again. This time, the robot actually rolled backward. It crested the hill on a subsequent attempt, but still, nothing like this sort of engine trouble had ever happened in testing.

Red Team had people stationed at designated viewing points that DARPA had set along the course. Reports came back that another engine stall likely happened fifty-four miles into the race. The stalls prevented the engine from turning a generator that created electricity for the sensors. Backup batteries were able to provide some power, but not enough for the main LIDAR unit. That was set in a gimbal, which a helicopter camera crew revealed was positioned at a ninety-degree angle to the direction of the robot's travel, rendering it completely ineffective.

The disabled robot slowed so much that the second entry to leave the chute, Stanford's Stanley, caught up to H1ghlander at mile 73.5. DARPA had promised its contestants that their robots would be navigating a static environment, meaning nothing could move in any of the contestants' fields of view. To prevent Stanley and H1ghlander from confusing each other, DARPA used a radio transmitter to "pause" Stanley for 2 minutes and 45 seconds, allowing H1ghlander to go ahead, creating some territory between the two robots. But soon after Stanley was reactivated, the robot caught up to H1ghlander a second time. This time DARPA paused Stanley for 6 minutes and 35 seconds. But Stanley caught up to H1ghlander a third time. Finally, at mile 101.5, 5 hours, 24 minutes and 45 seconds into the race, DARPA paused *H1ghlander* and allowed Stanley to take the lead. "Stanley has passed H1ghlander," Tether announced in the observation tent, prompting Thrun to leap into the air in triumph.

Shortly after, and with an elapsed time of 6 hours, 53 minutes and 58 seconds, Stanley became the first robot ever to autonomously complete a DARPA Grand Challenge. Tether himself waved the checkered flag as Stanley passed the finish line.

Sandstorm launched at about 6:50 A.M. The robot rumbled out

of its chute with its characteristic diesel knock. It made it through underpass one, two and three even though a software bug prevented the LIDAR from detecting the walls. In fact, it performed flawlessly until 6 hours and 30 minutes into the race, when it just scraped a canyon wall in the narrowest section of the route. Sandstorm drove over the finish line 7 hours and 4 minutes after it left the start chute—a variation of only about 1 percent from what the engineers had asked of the robot. It had done exactly what it was assigned to do in remarkable fashion, placing second, by time. And in third, limping into the finish, was H1ghlander, with an elapsed time of 7 hours and 14 minutes, or 55 minutes longer than the time the Red Team had set for it. All told, five robots finished the course.

Thrun was elated, of course. Later that day he and his team gathered onstage to receive a check for $2 million. But what was just as gratifying was the way the victory felt like a validation of the whole robotics field. More than a decade later, public attitudes toward roboticists have markedly changed. Back in 2005, robotics was associated in the public imagination with projects like Thrun's 1998 Minerva museum tour guide—as novelties, curiosities that had little effect on anyone's day-to-day lives. A self-driving car was different. Sure, the second DARPA Grand Challenge was a controlled scenario separate from the actual world because nothing else on the course was allowed to move. But it nevertheless represented a step toward actual robot cars, which everyone realized would, if they ever became a reality, transform lives. Standing up before reporters frantically scribbling down their words, photographers and videographers capturing their images and a crowd of people cheering their accomplishment, Thrun and his teammates relished the attention as validation that the world might finally recognize the potential of their chosen field.

Thrun was magnanimous in his victory. "It's really been us as a field that were able to develop these five vehicles that finished the race," Thrun said. "It's really been a victory for all of us."

Few on Red Team felt that way. It stung that they had devoted months to test Sandstorm and H1ghlander on some of the toughest roads on the planet—and then discovered on race day that the course was easier even than the well-graded roads that had marked the first Grand Challenge. It stung, too, that based on its performance in the qualifying events, a fully functioning H1ghlander would have taken the race. And it stung that, had Red Team's leadership allowed Sandstorm to perform to its abilities, rather than playing it safe and limiting its speed, the older Red Team robot also might have beaten Stanley. Thrun acknowledged both facts. "It was a complete act of randomness that Stanley actually won," he said later. "It was really a failure of Carnegie Mellon's engine that made us win, no more and no less than that."

"It was very much a winner-take-all event," Urmson recalls, more than a decade later. "It sucked. There was no prize for second. This had been three years of people's lives at this point. It was brutal. I remember seeing Red afterward, and that was the most distraught I'd ever seen him."

"It's right up there with the worst shortcomings of one's life," Red says, assuming full responsibility for what he still regards as a defeat. "I let a team down. I let a lot of people down. And in a lot of ways, in a bigger way, I let down a community and a world that didn't see the best of the technology and the movement and the vision of what things could be."

"It was a strange feeling," Urmson says. "It was a day that five vehicles did something believed to be impossible. Our team had pulled together and achieved the impossible. We'd done the impossible—and yet we'd lost."

Chapter Three

HISTORY HAPPENS IN VICTORVILLE

An introverted engineer looks at his shoes when he talks to you.
An extroverted engineer looks at your shoes.

——UNKNOWN

The second DARPA Grand Challenge was successful on numerous fronts. The $2 million prize was perceived as a cost-effective way to spur progress in the field of mobile robotics. The large number of entrants, the public enthusiasm and media attention for the challenge event, and the fact that the race resulted in five vehicles that could travel 132 miles through a difficult desert landscape all contributed to a perception throughout the military of money well spent.

But inside DARPA, there remained a sense that the mission hadn't yet been accomplished. No team had constructed a robot that could navigate the chaotic urban environments of Iraq or Afghanistan. Could a similar event spur the robotics field to make more progress?

Thus came the idea of the DARPA Urban Challenge, which would be staged in a *city* landscape, rather than a desert. Tether announced the event in April 2006, setting the date for November 3, 2007, and soon saw a globally diverse field of eighty-nine teams registered to compete, less than half the number of the previous contest, perhaps because this version of the challenge was perceived to be much more difficult.

Some of the format changes seemed designed to stifle the approach that Red Whittaker's team had pioneered in the first and second races—the one that had used a group of human map techs to essentially pre-drive the race route for the robots. DARPA planned to sprinkle the course with moving obstacles, namely, other automobiles driven by Hollywood stuntmen and professional drivers. As well, many different teams would be navigating the urban environment simultaneously.

The key to this event was satisfying the objectives DARPA wanted the robots to execute throughout the course—and DARPA wouldn't disclose that information until five minutes before start time. In fact, DARPA was so secretive in the lead-up that for a time they even declined to disclose which state would host the competition. "We knew it was going to be cold outside, so they'd probably do it somewhere that had warm weather," recalls one participant. "But that was the only thing we knew . . . They didn't want anyone pre-programming the system. They wanted some element of intelligence and route planning and control in the robot."

The rules required the robots to drive sixty miles in six hours through an urban environment while obeying the rules of the California Driver Handbook. The challenge would require that the robot navigate a standard North American parking lot well enough to be able to maneuver into an available space. While neither pedestrians nor cyclists would be allowed on the course, the robots would have to navigate one of the most difficult elements of driving for human operators—deciding how to proceed through an all-way-stop intersection where other drivers have arrived at about the same time.

"The Urban Challenge was much harder, in terms of what the vehicle needed to do," recalls Urmson. "The algorithmic steps we'd taken for the first two challenges were predicated on the world not moving. Once other things start moving, it's nowhere near as easy."

The inspiration here was to automate the operation of battlefield convoys. A military truck in Afghanistan or Iraq is transporting food to a distant village. An IED explodes somewhere ahead, and the automated convoy would have to navigate around the disturbance without running into any medics, civilians or other members of the convoy. That's about as dynamic an environment as is possible.

There was no question that Carnegie Mellon would enter the race. There *was* some question whether Red would lead it. Previously, Whittaker's Red Team had been a take-all-comers effort populated by undergraduates, volunteers, grad students or the odd full-time employee of Whittaker's Field Robotics Center. But this time around, DARPA was pledging a million dollars of research funding to a selection of the best-run teams. Carnegie Mellon was one of the recipients. As well, the stakes felt higher this time. There was the $2 million prize. In addition, Carnegie Mellon was competing for its reputation as the nation's top robotics center. It *needed* to win. "That's a lot of money and so [the university administration] wanted to make sure we could win it," Urmson recalls. Ultimately, Whittaker retained his leadership of the Urban Challenge team, but this incarnation featured other senior members of the Robotics Institute faculty—the university's equivalent of a supergroup.

Some in the university felt that this new group was so different it deserved a different name. Red Team had been appropriate in the past because it had been Red Whittaker's baby. But this Urban Challenge was the best squad that Carnegie Mellon could assemble. Some of the veteran members of Red Team resented the rebranding. "It made no sense to us," recalls Michele Gittleman, Red's assistant at the time. "Everyone knew Red Team. The brand had already been established. We had hats, T-shirts, jackets." Nevertheless, to reflect the sense that this was a new effort, fully backed by the university, the Carnegie Mellon team rebranded itself as Tartan Racing, in a nod to the nickname given the university's sports teams, itself a reference to founder Andrew Carnegie's Scottish heritage.

This latest effort was going to require a lot more than a million dollars to fund. So in 2006, Whittaker came and visited me at the General Motors Technical Center. "Why do you think you're going to win this challenge?" I asked him. "Dust," Whittaker replied, explaining that the robots created by many other teams treated dust clouds as impermeable obstacles, unable to be driven through, while Carnegie Mellon's software and sensors were able to correctly tell that dust represented no obstacle at all. Although dust represented a comparatively small factor at the actual Urban Challenge, the answer convinced me. I liked Whittaker from the moment I met him. With his military bearing and his eminently American positivity, his confidence that ingenuity and hard work could solve any problem if you persevered, he struck me as a throwback to an earlier era of technical innovators, the sort of people who pioneered the automobile a hundred years before. I arranged for GM to back Tartan Racing in numerous ways. The corporation would end up providing Whittaker's team with $2 million in support, making us the team's lead sponsor. We also provided the services of some of our top engineers, and embedded one of them, Wende Zhang, with the Tartan Racing team in Pittsburgh. Tartan Racing's vehicle would become a 2007 Chevy Tahoe dubbed "Boss," after my long-ago predecessor as General Motors' vice president of research and development, Charles "Boss" Kettering. Other sponsors also donating funding included the construction-equipment maker Caterpillar; the auto-parts supplier Continental; and Applanix, a manufacturer of GPS systems.

One of the new personalities on the Tartan Racing team was Bryan Salesky, who ran the software team. Salesky had been working at a Carnegie Mellon Robotics Institute spin-off called the National Robotics Engineering Center. Whittaker cofounded it in 1994 with $2.5 million in funding from NASA, to commercialize the technology created at CMU's robotics department. The place is located in a nineteenth-century former iron foundry in the Lawrenceville area of Pittsburgh, on the shores of the Alleghany River. Its job was to partner with such companies as John Deere and Caterpillar to develop commercial projects like a self-driving harvester or an autonomous excavator. Shortly before work began on the Urban Challenge, Salesky was working on an autonomous navigation system for the U.S. Army. But the project wasn't going anywhere, and

Salesky was becoming frustrated with the slow rate of progress that he thought endemic to working directly with the government.

Just twenty-six years old, Salesky was, on the surface, a bit of a strange fit for such a senior position in Whittaker's crew. Guys like Spencer Spiker, Kevin Peterson and Chris Urmson are equally at ease with welding torches and air impact wrenches as LIDAR sensors. Salesky is more at home in a computer lab than a mechanic's shop. But he hit it off with Whittaker's guys, particularly Urmson, likely because the two men shared a Midwestern distaste for pretense.

In 2006, as teams across the country geared up for the Urban Challenge, other figures who would become main characters in the mobility disruption began bumping up against one another. For example, Dave Hall and Anthony Levandowski. Hall, then fifty-five, was an inveterate tinkerer and self-trained inventor. He first gained notoriety in the world of high-end audio, having famously built his first amplifier at the age of four. Living then in Connecticut in a family with lots of technical heritage—Hall's father designed atomic power plants, and his grandfather was a physicist—he already knew how to read electronics schematics when he went to college for mechanical engineering. While there he invented a version of a tachometer, a device that measured the rotational speed of things like wheels and propellers. Licensing the patent from that invention gave him enough income that, out of college, rather than getting a job, Hall moved to Boston to set up his own little technical shop. For a time, he lived by monitoring the world of government research contracts and building prototypes for the big defense firms. In the seventies, Hall invented a subwoofer, a type of speaker that provided clearer bass tones in stereo systems. With $250,000 lent to him by his grandfather, Hall moved to California and set up a company with his brother-in-law to manufacture the subwoofer.

That was in 1979. By the turn of the millennium Hall's company had sixty employees and a few million dollars in annual sales. Hall made a good living, but he was bored. Building remote-controlled robots that competed to "kill" other machines on *BattleBots*, the television show, occupied him for a time. (One of his constructions, Drillzilla, relied on the blade from a circular saw as a weapon.) He competed in the first DARPA Grand Challenge with a Toyota Tundra entry notable for its use of a stereoscopic camera setup, rather than LIDAR, to sense the road. In the interim between the first and second races, Hall became fascinated with LIDAR's potential. Tinkering as he was wont to do, Hall determined a way to cram sixty-four lasers on a single device, more than anyone ever had before. But the big innovation was that Hall's LIDAR *spun*. Previous laser range finders had stayed static, shooting lasers to get a limited field of view, a little like the way a human sees the world only in front of his or her eyes. By mounting the device atop the vehicle and engineering the LIDAR so that it revolved ten times a second and 360 degrees around, Hall's mechanism provided a complete portrait of the area around the vehicle. The new technology in effect created a three-dimensional scan of the world. It helped Hall's robot progress along the second Grand Challenge route faster than most of its competitors, although a mechanical failure prevented it from finishing the race.

Once DARPA announced the Urban Challenge, Hall saw that his LIDAR would be even more valuable, because the 360-degree field of view that his device provided would help the robot detect oncoming vehicles in all directions. So he set up a manufacturing operation within his subwoofer company, Velodyne, and hired as a salesman one of the brightest minds from the first two races. The person he hired was Anthony Levandowski, the thin, six-foot-seven-inch UC Berkeley grad student behind the first challenge's Ghost Rider motorcycle.

Which is how Levandowski came to be in Pittsburgh at the old Coke Works test facility—now referred to by most as Robot City—one day in late 2006, about a year before the race. Tartan Racing had bought one of Velodyne's LIDARs—a significant investment at about $75,000 a pop—and Levandowski flew out to Pennsylvania to help Urmson and team install it. He set it alongside the other sensors that sat on the metal latticework on the roof of Boss, the Chevy Tahoe. Levandowski anchored down the sensor, activated it, and the numerous computer scientists and engineers who had gathered to witness this moment watched as the device generated the rotational momentum it required to work properly. On a nearby computer screen appeared the ghostly dot matrix that formed the device's output. It was impressive—a million data points that could recognize up to 120 yards away everything from parking curbs to people's faces.

Then something came loose on the LIDAR, and the rotational momentum flung a counterweight across the room—hard. Only luck prevented the wayward schrapnel from injuring someone. There was a shocked silence, and then came Levandowski's voice.

"We'll fix that," said the gangly engineer.

Levandowski would become well known years later as the central figure in an important lawsuit between Waymo and Uber. Brilliant and ambitious, Levandowski demonstrated a tendency to get himself into situations others would characterize as conflicts of interest. He was often in demand at the smartest, highest-performing groups working on the most intriguing projects. Velodyne would ultimately sell its LIDAR to at least seven of the Urban Challenge competitors, including two each to the Carnegie Mellon and Stanford teams. Around the same time Levandowski was selling this integral piece of technology to as many teams as he could, however, he also was advising the Stanford team for the Urban Challenge. There's no indication he did anything unethical at this point; Stanford apparently knew about his role with Velodyne, but it is easy to see how this activity could be perceived as a conflict of interest. And just to make all this even more incestuous, Google's Street View played a key role in how it all went down.

By 2006, Thrun was itching to do a start-up. The atmosphere of Silicon Valley played a role, as did the Stanford AI prof's developing relationship with Google cofounders Larry Page and Sergey Brin. But what should the start-up do?

Thrun was fascinated with the data that he and Montemerlo had collected while testing Stanley for the second Grand Challenge. Teaching Stanley to drive, motoring in the Touareg around the Mojave Desert, Thrun and his teammates had struck on the idea of fixing cameras, pointed in several different directions, on the roof of the vehicle. Not as sensors—rather, the imagery helped them recreate the circumstances that triggered bugs in Stanley's programming. Over time, Thrun realized how interesting it was to just go through the imagery collected by the multidirectional rooftop cameras.

That year, Thrun happened to be teaching a class at Stanford about computer vision. He assigned the class's most brilliant student, Joakim Arfvidsson, the task of creating a program that could easily stitch together the camera footage in such a manner that it provided the illusion of an unlimited field of vision, which, Thrun figured, would provide the impression of actually being at the original location. "Joakim recorded a street in San Francisco," says Thrun. The resulting computer program gave the feeling of actually standing on the street. "You could look up, look down—it was completely amazing."

Thrun gave Arfvidsson an A-plus in the class. During the summer of 2006, alongside his oversight of Stanford's Urban Challenge team, Thrun assigned a second team the task of building a version of the street-visualization software that could work on a cellular phone. The team included Hendrik Dahlkamp, Andrew Lookingbill and Arfvidsson. Thrun installed his good friend Astro Teller, whom he knew because Teller did his PhD in artificial intelligence at Carnegie Mellon, as the start-up's CEO.

In the first months of 2007, Levandowski also joined the team. Preparing to show the technology to venture capital firms for seed funding, Thrun aimed to stage a really impressive demo—he wanted to be able to place the VCs on any street in San Francisco, which required driving their camera rig along every street in the city. Levandowski was the one who determined how to do that quickly. He figured out that rental cars were really cheap if you hired them by the month. Next, he procured large numbers of drivers by advertising on craigslist. It took about two weeks to complete the map.

"He was a really great person, I would say, to get shit done," Thrun recalls.

March 2007 was the month Thrun made his approach to VCs for funding. The effort became all-consuming. "Just strategizing with these venture capitalists becomes this amazing, intense game," recalls Thrun. "My life is completely taken over. I have no social life—my wife thinks I'm a moron." The effort paid off, though. Two of the valley's top VCs, Sequoia Capital and Benchmark, were both interested. Thrun set the bidding for a Sunday, April 8, 2007. Soon the bids were climbing: a $5 million round of seed funding, which became $10 million. Then $15 million.

That evening, as he mulled over which venture capital firm to choose, Thrun invited himself over for dinner to Larry Page's place. Sergey Brin showed up. The men discussed Thrun's technology, which he called VueTool. Page already had sponsored something similar. Some time before, he, Brin and Marissa Mayer had gone out and taken some footage that they then stitched together. (In fact, according to Thrun, the idea of immersively stitching together camera imagery so that it was possible to click through it had been invented in 1979 by an MIT scientist named Andrew Lippman.) And Page and Brin had a similar project happening at Google, one run by a guy named Chris Uhlik. After dinner, Thrun took Brin and Page to his office at Stanford to demo the footage of the San Francisco streets. The Google cofounders were impressed; they saw that Thrun's team had accomplished a lot more, much more cheaply and in a lot less time, than their people had. For example, Google's in-house Street View team was using custom-built camera rigs that cost $250,000 each, according to Mark Harris's reporting in *Wired*. Thrun and Levandowski, Harris writes, were getting images of similar quality using a setup of off-the-shelf panoramic webcams that cost $15,000.

The next day, Google's head of mergers and acquisitions called Thrun, who agreed to sell the VueTool technology to Google. As part of the deal, Thrun, Levandowski and the rest of the team joined the company to accelerate the Street View project. "We got fairly lavish bonuses," Thrun explains. In following with the Red Whittaker approach of setting an ambitious goal to motivate his team, Thrun set up an arrangement with Google that would trigger another bonus payout if Thrun, Levandowski and the team were able to map a million unique road miles for Street View. Using the method that Levandowski pioneered, with lots of cars, Thrun and his team ended up meeting the milestone in just seven months.

Thrun's obsession with meeting the Street View goal meant the day-to-day work on Stanford's entry to the DARPA Urban Challenge was led by Mike Montemerlo. On the other side of the country, Montemerlo's former officemate Chris Urmson led the day-to-day work on Carnegie Mellon's robot vehicle. Urmson approached his effort on this race like it was the most important quest of his life. Every so often he recalled the conversation with his wife, Jennifer, nearly four years before, when he promised that he'd just do the first desert challenge, before going off and getting a real job outside academia, where he could start making a good living to support his growing family. (He and Jennifer had since had a second boy.) Sandstorm's rollover before the first race had provided him with the sense that he could have won it, if not for that accident. Highlander's mystifying mechanical issues in the second DARPA challenge provided a similar sense of the team's tantalizing proximity to victory. This urban event, Urmson knew, likely represented his last chance at victory. His last, best shot.

Leading up to the race, Team Tartan often discussed the rules. How difficult would DARPA make this challenge? Did DARPA even *want* a winner? It would be a simple enough matter to make the race so tough that no team could ever win. That would be the most cost-efficient option. If the government's intention was to outsource development of autonomous vehicles, to prove the possibility of self-driving cars while investing the minimum amount of money, then one way to do that would be to stage a race that prompted universities and research centers all over the country to work on the problem, while making it so tough that DARPA wouldn't have to actually hand out the prize money.

By this point, building an autonomous robot had become nearly routine. Transforming a Chevy Tahoe into the self-driving Boss resembled the maturation of a human being, in some respects. The vehicle started blind and dumb, unable to sense, to navigate, to move on its own. Then Urmson and his team installed sensors—the LIDAR, the radar—as well as the computer processors. In their earliest tests the team taught the robot not to walk but to drive, supplying it with a list of GPS waypoints, similar to the sort that earlier generations of CMU vehicles had been required to follow in the desert challenges. Once the robot could trace the dots of a mile lap around the grounds of the old steel mill, the team set up longer waypoint-finding tests. In November 2006, a full year before the competition, Boss completed a fifty-mile route, achieving speeds of 28 mph.

In parallel to Boss's mechanical testing, Salesky's programming team worked to incorporate perception and planning systems into Boss. The robot could understand the input it was getting from its rudimentary eyes, in the form of its Velodyne LIDAR and radar sensors. In December, Boss achieved a multi-checkpoint mission, running at night along the cold Pittsburgh riverside. Tartan Racing also coded rules that would dictate how the vehicle would behave during the situations it encountered while driving. Intersection handling was an early module in which the programmers created a set of directions for the various situations the robot might encounter at an all-way stop. What if Boss arrived first at an intersection, followed by someone else to the right? What if Boss arrived second? Salesky's team created rules for each scenario.

Around the same time, Tartan Racing integrated its behavioral software with the hardware. A synchronization board regulated the timing of the information coming in from each disparate sensor, allowing Boss's computing cluster to build a simulation of 3-D reality, the same way human drivers use their eyes and ears to build a model of the world in their heads. Another step involved predicting the behavior of other objects. For cars to be truly autonomous, they would have to be able to anticipate the behavior of practitioners of many different modes of urban transportation, from pedestrians to cyclists, skateboarders and scooter riders, among many others. But DARPA had told its teams that this race would happen in a vastly simplified environment; the only moving objects on the suburban course would be other automobiles. That made things a lot easier for the men and women coding the software, because it meant that Boss had to understand only a single set of behaviors—a tendency to move only forward and back along curvilinear lines, for example. Everything else, to Boss, was a stationary object.

As 2006 gave way to 2007, I flew to Pittsburgh to visit Red, Urmson, Salesky and the rest of the Tartan Racing team members at Robot City, including the GM engineers working with the team. I can remember touring the team's workspace in the winter, amid these post-industrial ruins, the railroad roundhouse, the trailer set on the frozen ground, and being struck by how chilly the offices were. Everybody walked around in winter coats with knit caps on their heads. You could see your breath. These guys are really living lean, I thought. Rather than spending their budget on creature comforts, it impressed me that they were reserving it for research and technology.

Despite the Spartan feel, Tartan Racing's headquarters seemed glamorous to me. I was a top executive in a major corporation, managing a budget in the billions of dollars at GM, but a part of me envied these young men. They weren't the sort of people who crunched numbers in an ivory tower. They were "salt of the earth, roll in the mud, get your fingers dirty" engineers—who just happened to believe they could change the world. Shielded from the university bureaucracy by Whittaker, unencumbered by the sort of corporate red tape required by General Motors, these guys were getting things done that GM never could.

I watched as Whittaker, Urmson and the Tartan Racing team tested Boss in the sort of situations that had been modeled using computer simulations only weeks before. Boss's first challenge at the demo I witnessed was a three-way intersection. The rules the programmers had coded into Boss's digital brain guided the robot to yield to other vehicles that had arrived before it, and when it was Boss's turn, the robot rolled through without a problem—an enormous relief to the team. As an additional feature of complexity, the team incorporated bad drivers into the equation. In one such test conducted at Robot City, Boss arrived at an intersection after a white American-model sedan, followed by a third vehicle—the second Humvee donated by AM General before the second Grand Challenge. The white sedan went first. Boss inched forward—and then the Humvee sped through the intersection, completely out of turn. Rather than rolling into the Humvee, Boss paused—exactly the appropriate action.

I was so excited I asked to go for a ride in Boss, which earned me some looks. The team didn't seem certain it was a good idea. But I insisted, and moments later, I was squeezing myself into the only bit of interior cockpit space not occupied by computer equipment or batteries. Soon I grasped how different this robot was from the sort of vehicles my team designed at GM. Boss accelerated toward a stop sign and braked at the last minute. It careened around a turn and spun its tires over potholes and rocks without slowing. Every motion happened with a jerk or a slam. Just a minute or two into the ride I felt carsick for the first time in years. Now I understood their hesitation. Boss wasn't designed for human occupancy, Whittaker explained soon after my ride. Rather, it had been designed for the express purpose of winning DARPA's third race. The robot had been programmed to accelerate aggressively, and brake hard when the situation warranted. The herky-jerky behavior made humans motion sick—but it also made Boss fast.

After an open house inviting the public to witness a demonstration of Boss's autonomous driving, the Tartan team loaded Boss into a tractor trailer and drove it to Arizona, where my staff had arranged for the robot to be tested at the GM Proving Grounds at Mesa. The warmer weather and the wide-open space of GM's test tracks allowed Boss to begin learning how to behave in parking lots—which DARPA had said would be a key part of the Urban Challenge. The team modeled left-hand turns into traffic, a difficult proposition for many human drivers. And when the temperatures warmed up in the spring, Boss returned to Pittsburgh to prepare for a visit from DARPA, during which the Urban Challenge's program manager, Norm Whitaker (no relation to Red), would monitor the sentient Chevy Tahoe as the robot conducted a series of tests. How the robot performed would dictate whether Tartan Racing would be one of the teams that remained in the race when DARPA staged the next round of eliminations. The resulting shortlist would compete at the so-called semifinals, the national qualifying event at the end of October 2007.

During the site visit, Boss had to pass four different tests. These challenges were conducted on Robot City's quarter-mile track. Norm Whitaker and a crowd of about a hundred, including media and sponsors, watched as Urmson and his team demonstrated that Boss's emergency stop button could halt the vehicle at a moment's notice. The robot had to pass through an intersection also used by other vehicles without a collision, which it did. Another challenge involved driving down a street and avoiding a parked car—no problem. "Boss behaved like a good beginning driver," praised DARPA's Norm Whitaker. "A real good job."

At the beginning of August, DARPA director Tony Tether announced the names of the thirty-five finalists invited to compete at the national qualifying event in late October. Tartan Racing was among them, which Whittaker, Urmson and Salesky were expecting. What they didn't expect was Tether telling a reporter that Boss was not considered to be among the top-five robots.

Consequently, the final months before the qualifiers saw Urmson and Saleskey conducting exhaustive testing of Boss to ferret out any bugs in the algorithms the programmers had constructed—to make Boss as safe and capable a driver as any human being. Some of this testing involved hiding an inflatable model of a car next to the road. Just as Boss was about to pass, a team member would shove the bubble car onto the pavement. Everyone watched to see whether Boss would react in time.

When the robot became expert at dealing with *that* situation, Urmson escalated the challenge, to using an actual car. One day in late summer the guys were testing in Arizona. Salesky was driving a rental car in front of Boss. "Okay," Urmson, who was riding shotgun inside the robot, said to Bryan over a walkie-talkie. "I want to make sure the vehicle's speed control is working. Slam the brakes."

The first time Salesky slammed the brakes, Boss slowed to a stop. But it wasn't abrupt enough for Urmson. "Slam them *harder*," Urmson told his friend. Moments later Salesky swerved in front of the robot then jammed his foot down on the brake pedal. "Harder," Urmson beseeched Salesky. "Make the tires *skid*."

This was Red Whittaker's modus operandi. It's only in the edge cases, Whittaker liked to say, the test-to-failure cases, that you learn about the robot's capabilities.

Salesky shrugged—and slammed on the brakes.

And Boss ran straight into the back of the rental car. Salesky climbed out and went around to look at the rear bumper. The back of the rental had crumpled like a spent piece of paper. The damage to Boss was worse, because many of its most sensitive instruments, including a pair of medium-range radar units, were bumper-mounted. "The collision blew up about ten grand worth of sensors," Salesky recalls. Urmson and Salesky stood there, arms akimbo, regarding the damage and shaking their heads. "Why did we just do this?" Salesky asked.

"My fault," Urmson said.

"No, I should have known not to do it that hard," Salesky said. He laughs about it, a decade later. "We both knew it was completely unrealistic. The following distance was way too close. There's moments when you're testing for so many hours in the field and you're not really thinking clearly—and that was one of those moments."

Sometimes, during such testing, a representative from DARPA dropped in on Tartan Racing to make sure that the team continued to make progress, to ensure that its money was being put to good use. "Hey," the DARPA rep asked Urmson and Salesky during one of these visits. "What are you guys doing to make sure you don't roll over your vehicles?"

Urmson froze at the question. It was kind of a ridiculous inquiry. The DARPA rep was just needling Tartan Racing's technical director. The Urban Challenge didn't feature any off-road territory, so Urmson wasn't testing Boss in the sort of environments that might lead to rollover accidents. There were no sharp turns in soft sand, for example, which had been the problem with Sandstorm. Nor did Boss navigate arduous trails that featured the sort of ramps that flipped H1ghlander onto its side. "We will *not* roll over the vehicle," Urmson said confidently.

And they didn't. In fact, the worst accident was the rear-ending of Salesky's rental car. Urmson and Salesky spent weeks racking their brains to think up situations that might mess up the vehicle's computer algorithms, situations unlikely to happen in real life, let alone in the Urban Challenge. Times, for instance, that all four entrances to an all-way stop would feature the simultaneous arrival of vehicles. How would Boss handle that? Or they'd program in a route and then create an unexpected obstacle along the way, maybe staging an accident that completely blocked the road. Boss would have to calculate an alternate route. There were some near-misses, some last-minute swerves, but for the most part, Boss performed flawlessly. "That was a magical time," recalled Salesky.

Remember, this was 2007. No one had ever created a robot that could drive in traffic at the speeds that were commonplace on public roads. And as each day passed, Tartan Racing was realizing that it had done it. "The fact that it just generally turned on and worked . . . it *still* surprises me," recalls Salesky. "It was pretty amazing."

To demonstrate Boss's capabilities, Urmson posted videos on the Tartan Racing blog. One day, he got a phone call from his liaison with General Motors. The middle manager had been watching the videos Urmson was posting and he was alarmed—so alarmed, in fact, that the manager was considering suggesting GM pull its sponsorship. Urmson gulped. That would be disastrous for Tartan Racing, which needed GM's sponsorship money. It also would be an enormous black eye for the team.

This was a fascinating moment—one of the first examples of the disconnect between Detroit and the computer scientists and engineers who were working to make autonomous vehicles a reality. This disconnect would manifest later in a years-long rift between Detroit and Silicon Valley. Although the GM manager making the call would have worked under me, I only learned of this incident years later, via Urmson. I could see both sides. Throughout the 1990s, GM had developed a strong culture of workplace safety. Our testing protocols were extremely risk averse. The GM liaison was worried that Tartan Racing's testing was going to seriously injure someone. He also may have been trying to protect me, and my decision to sponsor the team. It would have been easy to interpret the near-misses in the videos Urmson posted as indications of serious issues with Boss's programming. GM's sponsorship of Tartan Racing represented a significant cash investment at a time of deepening financial instability. This was 2007, after all, right before the financial crisis. I was being forced to fight for every research dollar I could get. Anything but a win from Tartan Racing would be a serious personal embarrassment for me, and could compromise my credibility with my strategy-board peers.

Urmson realized he needed to conduct some stakeholder relations. "No, no," he said. "You guys don't understand—what we're testing is far beyond the scope of anything that Boss will face in the Urban Challenge. We're just trying to see what it can do—we think it's going to perform amazingly in the finals."

This reassured the GM middle manager. Heading into October 2007, Boss had eighteen sensors bolted, welded and glued to its exterior. The robot's luggage compartment housed 10 computers processing 300,000 lines of software code that could make driving decisions 20 times a second. Just months before, Boss could conduct its most complex maneuvers at 15 mph. Now Boss could conduct the same maneuvers at more than twice that speed. It could park itself in a busy parking lot. And if the road was blocked ahead of it, Boss could execute a three-point turn and plan a new route to its objective—completely autonomously.

All of which created an unusual amount of confidence for the Carnegie Mellon team, and for Urmson himself. The first race, Urmson was pretty sure they wouldn't finish. The second, he and Red and Peterson were so concerned about *no one* finishing that they set Sandstorm's speed too low and ended up taking second and third. During the 2007 site visits, Tether indicated he didn't think CMU even ranked in the top five. But something had happened to Boss in those final two months. This time, Urmson and the rest of the team felt as confident as they ever had heading into the national qualifying event.

That event began on October 25, 2007, in Victorville, California. The venue was George Air Force Base, which featured all the hallmarks of a regular town, such as buildings, roads, homes, apartment buildings and parking lots. All the hallmarks, that is, except for people, because George AFB had been shut down in 1992.

From the initial contestant pool of eighty-nine teams, DARPA judged that thirty-five had built robots that were good enough to attend the qualifiers. Arriving in Victorville was an exciting moment for each entry because it represented the first time that any of the team members could witness firsthand the competition. There were a lot of familiar faces at the welcome event that Tether and program director Norm Whitaker staged in a big tent on the first evening. After all, many of these contestants had competed in the previous challenges. There was Stanford and Tartan Racing, of course. MIT, Caltech, Cornell, and Princeton. Team Gray, robot enthusiasts who worked for an insurance company during the day, had come in fourth behind the Stanford and Red Teams in the previous Grand Challenge—and entered this race despite not receiving DARPA funding. And Team Oshkosh, the Wisconsin-based defense contractor, had shown up as usual with the most imposing entry of the bunch, TerraMax, based on a four-wheeled variant of the U.S. military's Medium Tactical Vehicle.

The qualifiers featured three different trials. Area A was full of densely moving traffic and required the vehicles to turn left onto a busy street that had human-operated vehicles going in either direction. Area B was a 2.8-mile loop that featured "the gauntlet," which required the robots to pick their way through a road made narrow by stationary parked cars on both sides. There were impromptu detours that would have challenged even human drivers. Another portion of this area required the entrants to find a parking spot in a lot populated by other cars—and to pull into and out of the spot safely, without hitting anything. (Area B didn't feature any other moving vehicles.) Finally, Area C featured other vehicles on the road as the entrant navigated a series of four-way stops. The final part of Area C gave the robot a spot to reach, then placed an unexpected roadblock before the goal, requiring the robot to stop before the roadblock, execute a three-point turn and plan a new route to the goal, completely autonomously.

DARPA assigned Tartan Racing to conduct the B qualifier first—the 2.8-mile obstacle course loop, which triggered a bug in Boss's programming. The Chevy Tahoe robot suddenly stopped its progress on the course and backed up, concerning DARPA race officials so much that they sent a radio signal to pause the vehicle's operation—not an auspicious start. Luckily for Tartan Racing, the pause effectively rebooted Boss, clearing the effects of the bug. Despite the problem, Boss performed better than any of the other robots that completed the B qualifier that first day.

Afterward, the Tartan Racing team headed over to Area C to watch the progress of Stanford, which they figured was their biggest rival. Stanford's 2006 Passat station wagon, nicknamed Junior, performed well at the intersections and executed U-turns nicely. But confronted with the blocked road on the way to a checkpoint,

Junior wasn't able to plan out a new route—something, the Tartan onlookers noted, Boss had been able to do for months now.

Boss's progress the next day, in Area A, featured its own problems. Area A was the one that required the robot to turn left onto a road crowded with human stunt drivers. The task provoked a lot of problems for other teams. Georgia Tech's robot didn't turn left at all; instead it collided straight into one of the jersey barriers and dented its front bumper. MIT's robot was too slow merging into traffic. The robot for the Golem Group, the UCLA-affiliated team that had funded its entry to the first race with the director's *Jeopardy!* winnings, had a feedback error that caused it to accelerate out of control over a parking block before DARPA activated the emergency stop.

Given these mishaps, Urmson and Salesky figured the area would be challenging for Boss. When Boss entered Area A, the robot handled the left turn into traffic without much problem. Then came a serious issue. Confronted with an oncoming vehicle, Boss stopped dead. For twenty seconds. Which in turn triggered something called error recovery mode.

As it happened, error recovery mode was one of the most interesting things about Boss. Much of the work for it had been done by the programming team of Chris Baker, John Dolan and Dave Ferguson. At these early stages of autonomous vehicle development, where the programmers were just learning how to calibrate LIDAR and radar sensors to help their robots perceive the world, a robot like Boss had a lot of similarities to someone who was inebriated. Sometimes, Boss's perception was off. The robot saw something that didn't necessarily reflect the real world. When a tipsy person sees double, he might turn his head to the right or left. He might shake his head, squint or widen his eyes. He might close his eyes, give his head a shake and open them again. Is he *still* seeing double? Maybe not. Maybe all that shaking, eye closing and squinting reset things so that his vision goes back to normal.

Baker, Dolan and Ferguson programmed Boss to do much the same thing in a part of the error recovery mode called "shake and shimmy." When Boss encountered a problematic scenario with low certainty, just like a drunk with double vision, it did things to attempt to reassess the situation. It might stop in place and then turn its front wheels back and forth. It might reverse a little bit and crawl forward. Or reverse, go forward and then turn a little bit. All these nudges and nods were intended to provide Boss with another look at the situation.

In this case, in Area A of the qualifiers, the problem was that Boss incorrectly thought its lane was too narrow to accommodate the width of the Chevy Tahoe vehicle. DARPA had provided its contestants with a digital map of the race course before the qualifying event. Similar to its approach in the first two races, the Carnegie Mellon team had gone over every aspect of the race routes and annotated the map to provide the robot with additional directions, in effect, pre-driving the most problematic areas. (Stanford's team did much the same thing.) As part of this annotation process, one of the team members had accidentally defined a lane as too narrow to accommodate two vehicles. The presence of the second vehicle convinced Boss that it didn't have enough room to pass—so it went into its "shake and shimmy" mode, slowly moving forward into the area it thought too narrow to navigate. In the inebriated analogy, Boss's slow forward progress was a bit like a drunk extending an arm to see whether a second door handle actually exists. Of course, it doesn't, and when Boss found out that it could proceed without a problem, it headed on through the challenge. Boss did well in Area C, too, but for a moment the SUV becoming momentarily stuck between a low-hanging tree branch and a cloud of dust that it apparently considered a solid obstacle. (The rare time that Boss, contrary to Whittaker's early assertion to me, couldn't handle dust.) "Shake and shimmy" mode saved the day there, too.

Stanford's Junior was after Boss in Area A. Curious to see how their rivals would do, much of Tartan Racing stuck around to watch. Something in Junior's software prompted the robot to stop on the back side of Area A and wait, completely still, for precious seconds before it mysteriously headed off again. Junior successfully managed to execute left-hand turns into traffic. But it was tentative when it did them, requiring much more than the ten-second window that DARPA allowed. Recall that Thrun had been concentrating on Street View for much of 2007, and that Montemerlo had overseen much of Junior's development. The Stanford robot's cautious approach likely reflected Montemerlo's risk-averse personal demeanor.

Coming off the qualifiers, Urmson and the other Tartan Racing members figured they'd done really well. Whether they were the *best* depended on how DARPA elected to score the event. What mattered more—safety or capability? If DARPA valued safety more, Junior's cautious strategy and painstaking left turns might carry the day. If speed, confidence and the shortest time through the course mattered more, then Boss would place far ahead of Junior. "The subjectivity of the competition is maddening," wrote Tartan Racing computer scientist John Dolan to his family in an email. "It's pretty obvious at this point that we will both be there on race day barring a catastrophic accident . . . If speed and ability to handle different situations are the determinants, we stand the best chance now of winning; if safety . . . is DARPA's watchword, we have some chinks in our armor [that] Stanford or perhaps another team may be able to exploit."

--- --- ---

The qualifiers featured their share of disasters. Team Jefferson, affiliated with the University of Virginia, was doing well in the Area C qualifying event until their robot ran smack into a railroad-crossing barrier—because the robot's sensors only detected objects that sat on the ground. Oshkosh's big TerraMax truck hit a vehicle in the parking lot and dragged it about eight feet before DARPA paused the robot. Another team's entry drove into an overhang that essentially decapitated its expensive Velodyne sensor.

Once Boss had run through each of the three areas, Tony Tether told Chris Urmson that Tartan Racing's qualifiers were over because Boss had performed well enough to make it to the finals. Meanwhile, DARPA asked Stanford to make a second attempt at the Area A qualifying event—the one with the left turn. Before they did, Stanford ratcheted up the aggressiveness of Junior's driving. All told, Junior would require three attempts before it passed Area A.

On the first of November, Tether gathered the remaining thirty-five teams to announce which of them had made it to the next and final round. All along, DARPA had prepped its competitors to expect that a maximum of twenty teams would perform well enough in the qualifying events to make it to the finals. "Well, twenty aren't going to be in the finals," Tether said on the microphone. "The number that are going is eleven."

Stanford, Oshkosh, MIT, Cornell and Virginia Tech were among the successful eleven. So were two teams from Germany, Philadelphia's Ben Franklin Racing, a team from the University of Central Florida, and a joint effort from Delphi, Ford and Honeywell called Intelligent Vehicle Systems. Finally, the eleventh team was Tartan Racing—and when Tether announced Whittaker's team, the DARPA chief noted that Boss was the top-ranked robot. The race was Carnegie Mellon's to lose.

— — —

I arrived at the Urban Challenge the following day to watch Boss compete. The spectacle was more impressive than I expected. Approximately three thousand media, team relatives and curiosity seekers traveled to Victorville to watch the challenge. Among them were Larry Page and Sergey Brin, who'd transported a planeload of some of their senior-most executives to cheer on the Stanford team, which they'd sponsored, and more importantly, to get a sense of the capability of this technology. I want to recognize that detail, because I think it's important. Compare Google's planeload of senior executives with the interest from Detroit. At GM, for example, we didn't send our CEO or, in fact, anyone else from the strategy board, except for me. I was the only senior executive from my company to attend the Urban Challenge, which demonstrates the comparative level of interest at the time. Google had a great sense of the capabilities of software and could see that the realization of self-driving cars was much closer than most people thought. Hardware-focused GM still thought of the technology as distant science fiction.

On the second of November, the day before the race, DARPA asked each of its eleven remaining teams to bring their robots to the starting chutes, one at a time, to perform a dry run of the race start. When Stanford's Junior launched, it headed out, no problem—then promptly did a U-turn and almost crashed back into the starting chute. Among the eleven teams, Boss was the only robot to have a completely problem-free and successful trial launch.

The starting chutes, little more than parking spots created by cement jersey barriers, formed a line across a vast expanse of tarmac. Because Boss was ranked number one among the competing robots, it had the honor of being the first to start the race, an advantage because fewer cars on the course meant fewer obstacles for the self-driving software to deal with.

The opening ceremonies were something to see. Helicopters whirled overhead. Somebody sang the national anthem as military on horseback paraded before the grandstand. Boss's chute was situated closest to the Jumbotron, which powered on to provide the audience with visuals of the contestants' progress. Finally, forty-seven helmeted professional drivers motored onto the course.

In the starting pit, Boss's rooftop LIDAR sensor commenced spinning. Salesky, wearing the blue vest that distinguished him as a competitor, looked up from a computer screen to see his idol, Apple cofounder and legendary programmer Steve Wozniak, tooling by on a Segway scooter. "Hey, guys," Woz said, giving Salesky a wave. But Salesky couldn't think about the fact that his idol had just acknowledged his existence. Instead, with mere minutes remaining before the 8:00 A.M. start time, Salesky and Urmson went through a checklist to confirm that every aspect of the robot was operating as intended. The guys were a little obsessive about this ritual, inspired by Levandowski's gaffe at the first DARPA Grand Challenge, when after eighteen months of work on his Ghost Rider motorcycle, he watched it tip over at the starting line because he'd simply forgotten to switch on the bike's most crucial piece of equipment—the gyro that kept it upright.

"Software version still correct?" Urmson asked.

"Yep," came the reply from Salesky.

"Engine is started?" Urmson asked.

"Check."

"Siren connected?"

"Check."

"The GPS is jammed," Urmson said. "What the hell?" He eyed the absence of any detectable GPS signal on the laptop computer hooked up to the robot vehicle.

The GPS was necessary for Boss to locate itself on the course. Without it, Boss was more likely to possess an inaccurate sense of its coordinates in space, which made it more likely to hit something. Boss needed GPS to work properly. If the team couldn't fix the GPS issue, it would be a disaster on the order of a rollover accident.

This sort of thing didn't happen. Not at yesterday's dry run of the start. Not anytime through the qualifying events. The Global Positioning System worked as a result of twenty-four satellites that orbited the earth at an altitude of approximately twelve thousand miles. At any one time, Boss's GPS modules should have been able to receive signals from numerous satellites.

Urmson figured the problem had something to do with the GPS module itself. He shouted out to the team's design lead, "Get the other GPS!"

The team member sprinted off to retrieve the spare module. Urmson informed the rest of the team of the problem, and within seconds a half dozen of them were scrambling over the SUV to examine every aspect of the GPS-receiving mechanism, from antennae to computer.

Salesky was not a hardware guy, so he didn't have much to do but sit there and think. "This car has been awesome the entire last two weeks," he reasoned. "In fact, in the last *month* it's been pristine."

"Chris," Salesky said. "This *can't* be a hardware issue."

Urmson looked at him. "But what else could it be?" Soon Urmson rushed over to one of the DARPA race officials in their fluorescent yellow vests. "It is highly unlikely that we will be ready to launch at eight A.M.," Urmson told the man, who radioed the information to his colleagues, Tony Tether among them. Soon every one of Boss's doors was open as the Tartan Racing pit crew crawled over the vehicle, attempting to identify the issue. A clutch of yellow- and blue-vested team staff and DARPA officials conferred at Boss's trunk.

"So we were happy, everything was all converged," Urmson said to the race officials. "Suddenly, bam, we've lost GPS, we've lost differentials. Three different receivers."

"None of the other teams are having GPS issues," one official noted.

Urmson's cell phone rang. It was his wife, Jennifer, who along with Urmson's mother and father and thousands of other onlookers was in the grandstands overlooking the race start, watching the hubbub in the Carnegie Mellon pole position and wondering what was going on—why hadn't they started? "What's up?" she asked, concerned for her husband.

"We don't know," Urmson replied, his voice tense. "We're working on it."

⸻

With Boss out of commission, DARPA moved to allow another team, Virginia Tech, to begin its race. Minutes later, with Tartan Racing no closer to solving its GPS issue, DARPA allowed Stanford's Junior to launch.

The sight of his rival's VW Passat pulling out of its chute made Urmson sick to his stomach. "We had worked so hard, and for so long, and we knew that we were pretty clearly the favorites," he recalls. And now, the young technical director who had thought he'd all but sewn up a historic win could feel his achievement slipping away from him.

I felt awful for the guys on the team. Boss and the race crew looked so lonely, a solitary vehicle in the middle of a vast expanse of tarmac, surrounded by an increasingly panicked handful of blue-vested Carnegie Mellon team members. Those of us in the grand-stands could sense that something major was wrong. No one knew what it was—least of all the Tartan Racing team.

Urmson gathered together a little knuckle of senior team members and race officials, including Bryan Salesky, Tony Tether and Norm Whitaker. "What's changed?" the men kept asking themselves, em-ploying a time-tested engineering strategy to get to the root of a problem. "What's different from before?"

Tether looked up and saw the enormous Jumbotron screen, which Boss, thanks to its number one position, was nearer to than any other robot.

"Hey," Tether exclaimed, pointing up. "Tell them to turn that off!"

Seconds after the Jumbotron went dark, Boss's GPS signals re-turned.

Was that it? Tartan Racing's pit crew watched, breathless, as the GPS signals stayed strong. Apparently the radio interference emit-ted by the Jumbotron had somehow affected Boss's ability to receive GPS signals. The whole of Tartan Racing exhaled for the first time since the pre-race checklist.

"Give us another minute or two to verify that it's good again," Urmson said to Tether, his tone relieved. "And thank you."

By the time Urmson and his crew resolved the problem, it was eight-thirty, and eight of the eleven teams had left their spots. The first-mover advantage had evaporated. But the resolution of the GPS issue left Urmson and his team near giddy with relief. "Now we get to see what Boss can do with traffic on the course, boys," Urmson exclaimed.

The grandstand erupted with cheers as Boss pulled out of its chute.

"That," Salesky would say later, "was the closest I ever came to wetting my pants."

<hr>

The actual race experience was trying for the Carnegie Mellon team. Because there couldn't be any radio contact with the vehicle, Tartan

Racing was not able to engineer a live feed that showed them what was happening. Some of the team had an obscured view as they sat in the grandstands. Others stayed in the pit and strained their ears to hear Boss's distinctive siren, which sounded as long as the robot kept moving about the course. (Each of the eleven remaining robots had its own, signature noise.) "I just kneeled there with my face in my hands for like six hours, just listening for that sound," Salesky recalls.

Boss encountered its share of problems. The Chevy Tahoe stopped as it moved from a dirt road to pavement, apparently interpreting the transition in the roads as an obstacle. The second major problem started when another robot in Boss's path turned into Boss's lane. Boss overreacted, swerving and braking at the same time to avoid the nearby competitor, and approaching a barrier wall that formed the road's boundary. Boss decided that it was too close to the barrier wall to maneuver safely and halted its forward progress.

In both cases, the "shake and shimmy" error-recovery mode succeeded in maneuvering the robot into a position that provided a new perspective, triggering a change of mind. In the first case, the transition from dirt road to pavement wasn't an obstacle and Boss decided it could just roll over the transition. In the second case, turning the front wheels back and forth moved Boss maybe an inch from the concrete barrier, just enough to convince the robot that it could now maneuver safely.

The third incident looked the strangest, and had the biggest potential impact on Carnegie Mellon's ability to win the race. Boss came up behind a vehicle that had just stopped at a stop sign. The Carnegie Mellon robot halted a few feet behind the first vehicle, same as a human driver might. But when the preceding robot moved forward, through the intersection, Boss stayed still. Seconds ticked by. Abruptly, Boss performed a U-turn.

What was going on?

The team later worked out that it was a glitch in the software code of the planning system. Boss sensed the obstacle ahead in the form of the competing robot—but when the other robot left the intersection, Boss's software neglected to register that fact. It thought the obstacle was stationary and blocking the road. The U-turn happened because Boss had calculated an alternate route to the objective. The alternate route added an unnecessary 2.7 kilometers to the distance—but the important thing was, Boss got there.

The first vehicle to cross the finish line was Stanford's Junior, with an elapsed time to finish of 4 hours, 29 minutes and 28 seconds. The sight generated huge applause from the grandstands—sinking the spirits of pretty much everyone rooting for CMU, me included. Boss came in next. Then we realized that Boss had started much later than Junior. The elapsed time from the moment Boss began the race, to the time that it completed the course, actually turned out to be nearly 20 minutes faster than Junior. Boss's elapsed time was just 4 hours, 10 minutes and 20 seconds. But elapsed time to finish represented only one aspect of event performance. Each robot had to pass the California driving test. Finishing the event promptly didn't hurt one's chances, because it indicated an ability to efficiently plan and navigate the George Air Force Base course. Just as important, though, was the ability of the robot to follow the rules of the road: to avoid crossing the center line; to obey stoplights and stop signs; to prevent oneself from turning left at an intersection where such a turn is prohibited. *When* the robot came in from the course was important, but just as crucial was *how* the robot performed on the course during the race.

That night, Urmson, Salesky and the rest of the team went to bed knowing that Boss had performed well. They knew that they had a good chance to win. The next morning, all eleven teams gathered to watch as Tony Tether took the stage for the event's final ceremonies. While he climbed the stairs to the stage, Salesky thought about the project and the way it had taken over his life. The team's core members, guys like Urmson, Salesky, Peterson—they'd worked every single weekend for a year straight. "It was like 2007 didn't exist for me," Salesky recalls. "I looked back on it and my brother graduated that year from high school, my parents moved out of town—those are major life events that I barely remember. It was like, oh, okay, my brother's graduating? Great, congratulations, Zachary, anyway, I'm working."

The first thing Tether did onstage was attribute Boss's delay at the start of the race to interference from the Jumbotron's electromagnetic radiation. Although Boss came in after Stanford's Junior, Tether said, Boss's time on the Urban Challenge course was twenty to thirty minutes shorter than any other competitor's.

In the audience, Urmson allowed himself to breathe. Tether wouldn't be taking the time to explain all this if Boss hadn't come in first, would he?

"As you know, we have only three prizes to give out," Tether said. "But I do want you all to know that even those who aren't getting prizes—you're all winners. I mean, this is fantastic."

Next, Tether explained the way the judges evaluated each of the robots, which included using video footage from the helicopters that had been flying overhead throughout the race. None of the top three finishers had committed any major safety incidents, Tether added.

Urmson's spirits soared at that information. If neither Boss nor Junior had committed any safety violations, then the fastest robot would finish *first*, he reasoned.

Third place, Tether announced, went to Virginia Tech, and the team leader walked onstage to receive from Norm Whitaker a supersize version of a check for $500,000.

The next check was covered in brown craft wrapping paper. "Who got the million?" Tether shouted, teasing the crowd before he disclosed the second-place finisher. "Junior!" Tether shouted, pulling off the craft paper to reveal the check.

Which meant that Boss, and Carnegie Mellon, was the winner. More than triumphant, Urmson, Whittaker and Salesky were relieved. They'd done it. Finally, they'd won. I raced up to the stage and took a picture of the Tartan Racing team receiving their check.

Whittaker, Urmson, Salesky and all the rest of them, beaming as they took in their achievement.

"The real winner," Tether said when things had calmed down again, "was the technology." And he was right. That race was the turning point, where the future of connected, shared and driverless vehicles became possible. Later I asked Tether when the next race would be held. "There won't be one," the DARPA chief replied. "Mission accomplished."

It felt pretty good for all involved. On his way home from the event, at the Las Vegas airport, Red Whittaker was waiting for his flight when a teenager saw the DARPA Urban Challenge hat he was wearing. The kid asked, "Did you take part in that race?"

"I *won* that race," Whittaker said.

Urmson was invited to appear with Boss on the *Today* show, so that television viewers could get a look at the vehicle that won the first-ever urban driving challenge. A little later, at the beginning of 2008, I took Boss to Las Vegas to appear as part of GM's exhibit at the Consumer Electronics Show. I was enormously impressed by, and grateful to, Red, Urmson, Salesky and the rest of the team. So much so that I arranged for a special awards ceremony to present each team member with his or her own replica of the bigger trophy DARPA had awarded to Carnegie Mellon. "Congratulations to the Tartan Racing Team, Carnegie Mellon University, and the other members of the team from GM, Continental, Caterpillar and our other partners for your outstanding achievement," I said. "You and Boss made history. What made this race so significant was that Boss and the other vehicles that competed represent a new DNA for the automobile—a new DNA that will ultimately replace today's vehicles."

I was thrilled for the team. I had gotten to know them well throughout the past year and grew to respect their abilities, character and passion for winning. I was also happy for GM. We were entering a period that would see us fighting for our very survival, and we could use any positive news to boost morale.

Back in Vegas, as GM staff demoed Boss to media, one of the reporters asked me how long I thought it would be until autonomous vehicles appeared on American roads.

"Ten years," I said, just coming up with a number off the top of my head.

The reporter thought I was crazy. But that was January 2008—and a little more than a decade later, my offhand prediction has turned out to be accurate.

II

THE NEW DNA OF THE AUTOMOBILE

Chapter Four

A FISH OUT OF WATER

Nonconformists are all alike.

—UNKNOWN

I was really excited when I returned to Detroit from Victorville. DARPA's three self-driving challenges created a community of engineers and computer scientists that would significantly advance autonomous-car technology. DARPA took a niche project pursued by the odd academic or hobbyist and set it on a path toward widespread societal transformation. Years from now, Tony Tether's decision to have DARPA invest in the three races—the federal agency spent approximately $9.8 million on the 2005 event, while the larger-scale Urban Challenge required the government agency to spend $25 million—will be lauded for its remarkable multiplier effects. This was smart government spending, because the races would contribute to the creation of a new form of transportation market. The investment spurred the development of a different way to get where you want, when you want, simultaneously safer, cheaper and more efficient, while minimizing environmental impacts and freeing up trillions of dollars of resources.

Driverless technology would not create this new market alone. Two other trends would play equally important parts. The electrification of motor vehicles would set the stage for cars that were easier to manufacture and didn't rely on oil for energy. The third trend was transportation as a service (Uber, Lyft, etc.), which set the stage for consumers to transition away from personal ownership of cars and toward ride-sharing services that allowed users to pay for their mobility in numerous ways: by trip, by mile, by monthly subscription or possibly a combination of all three.

Thanks to these converging forces, we're at an inflection point, a once-in-a-century opportunity to redefine not just the auto industry, but personal mobility itself.

Years from now, we'll regard as incredibly wasteful the way we got around in the twentieth and early twenty-first centuries. Future generations will encounter artistic depictions of contemporary living and they'll be shocked. Take the opening number in the 2016 musical *La La Land*, which begins with a tracking shot of a familiar scene to anyone who has driven in Los Angeles: four lanes of cars stopped dead for a stretch of freeway. The bored drivers lean out their windows, listen to the radio, look curiously into other cars—and then a woman begins singing a song. She steps from her car out on the freeway pavement to dance. Soon the drivers of the other vehicles are joining her, and suddenly it's a raucous party. Future generations will see the scene and they'll take in the kettle drummers in the cube truck, the BMX bikers and the skateboarders, the catchy song and the kinetic choreography, but what may astonish them the most is the mere fact that people are stopped on a highway at all. "What's a traffic jam?" they'll ask, never having experienced one before because the robot cars of the future will manage traffic much more efficiently. "Why are those cars so big? And why are so many of those big cars only occupied by a single person?" And then maybe they'll use an app to summon a driverless two-seat electric car with a top speed of around 45 mph, which will arrive in a minute or two to bring them to any destination they desire.

I had my own role in this disruption. While the Carnegie Mellon and Stanford teams were working on the challenges, I was advocating for a more rational and sustainable approach to automobile transportation as General Motors' corporate vice president of research, development and planning. I was motivated by the statistics I mentioned in the introduction—personally owned vehicles that sit idle 95 percent of the time, for example. When used, those vehicles transport a single occupant 70 percent of the time, in most cases propelled by fossil fuel, from which just 1 percent of the energy goes to move the driver. The system is completely irrational. To understand how we arrived at this situation, it's necessary to look at the past.

Nikolaus Otto of Germany invented the four-stroke internal combustion engine in 1876, and his onetime employee Gottlieb Daimler succeeded in refining the device in 1885 until it was compact and powerful enough to propel a carriage. Elsewhere in Germany, Carl Benz created an automobile the same year. Charles and Frank Duryea, a pair of brothers, were the first to run a gas-powered automobile on American roads, in 1893 in Springfield, Massachusetts. Soon, inventors all over the United States were working on horseless carriages. According to Harold Evans's book *They Made America,* "more than 3,000 car companies were formed in America in the decade between 1895 and 1905, and hundreds of them actually put cars on the market."

The vehicles were expensive, at first, so they started as novelty items, the exclusive playthings of the wealthy, who delighted in zooming about and alarming pedestrians and the drivers of horse-drawn carriages. Consequently, the first cars were regarded by the emerging middle-class as egregious symbols of arrogance and wealth—something like the way today's drivers might regard a Lamborghini or McLaren supercar.

It's interesting to consider what might have happened had such vehicles stayed too expensive for the middle class. Mass transit might have become much bigger than it did. Ride sharing could have grown in popularity. Middle-class car ownership might have been restricted to taxicab owners, who might have invested large amounts to own a vehicle, driven most hours of the day, every day, moving people and goods around urban centers. If automobiles had stayed a luxury item, they might have been used much more efficiently.

But the fact is, Henry Ford came along and changed everything. Ford became one of the world's most famous and wealthiest men, occupying a place in American pop culture similar today to Warren Buffett, Elon Musk, and Bill Gates, all wrapped into one rail-thin package. In his satirical critique of industrial society, *Brave New World*, Aldous Huxley places Ford in the role of widely worshiped societal savoir, to the extent that his characters say "my ford" rather than "my lord" and have switched the phrase *anno domini* ("the year of our Lord") with *anno Ford* ("the year of our Ford"). In the same way that today's foremost tech companies often inspire neologisms (to "Google" something, for instance), the automobile mogul coined the term "Fordize" to describe both the process of mass production and the resulting decrease in the price of the mass-produced product.

Born to a farming family in 1863 just outside of Detroit, Henry Ford first saw a gasoline engine in 1890 at the age of twenty-seven. It happened while he was working as a steam-engine mechanic for Westinghouse in Detroit. On a service call at a factory, Ford saw a four-stroke motor known as an Otto, after its German inventor. Inspired, Ford sketched a design for an automobile built around an Otto, but he didn't pursue its development for several years. Ford completed his first vehicle at 4:00 A.M. on June 4, 1896; the shed in which he'd built the automobile had a door that was too small, so before he could actually go for a drive Ford had to demolish the doorframe with an ax.

For a time, Ford worked for companies focused on building one finely crafted and luxurious automobile at a time, unique to the customer's needs, for the highest price possible. But he had a different goal in mind: Ford wanted to create a mass-market automobile, each one of which would be the same as the next, at a price so low that almost anyone could afford one.

He incorporated the Ford Motor Company on June 16, 1903, and sold his first automobile, the Model A, to a Chicago dentist for $850. Business expanded quickly. One year in, Ford employed 125 people and had sold a thousand cars. In 1907, Ford gathered his top engineers and challenged them to design a burly vehicle, strong enough to carry five people and yet light enough to reach high speeds. With a new-to–North America vanadium steel, a 20-horsepower engine, a 100-inch wheelbase and tall tires that could roll across the American countryside, where just 20 percent of roads were paved, the first production vehicle of the Model T rolled out of Ford's Detroit factory on September 27, 1908.

They built just 11 vehicles the next month, but production quickly ramped up, with Ford assembling more than 10,000 in 1909. From there, he refined his assembly methods. He borrowed ideas pioneered by the Chicago meatpacking industry, which processed cows and other animals by directing stationary butchers to remove the appropriate part of the animal as it passed, suspended from a hook that rolled along an elevated track. Ford's application of the idea to build the Model T started with a chassis that rolled along an assembly line through a factory, where stationary workers bolted, welded, sewed and glued on successive parts until the result was a complete Model T. The company refined its innovation until by 1914 the assembly of a Model T required just ninety-three minutes, start to finish. The price of the Model T dropped as Ford increased his sales. Five years after the first complete year of production, half the cars on American roads were Fords, helped in part because the Model T was listed at just $440— about $11,000 in today's dollars—half its initial selling price.

Attendant to the idea of mass production is mass *consumption*. Ford's dream, according to his autobiography, was to build a car "so low in price that no man making a good salary will be unable to own one." I've always found it interesting the way the phrasing of Ford's dream portrayed car ownership as something mandatory. He helped to create an America where having a vehicle was not only desired—in some respects, it seemed *required*, one to each American. Was this rational? Did it make sense to spend a significant proportion of your income on a five-passenger vehicle that you alone might use just thirty minutes a day, creeping along in traffic most of the time? When they were as cheap as the Ford Model T, who cared?

The other innovation that Ford ushered in, which helped bring us to a present of more than one car per driving American, came about as a direct result of Henry Ford's mania for efficiency. The factories that created the Model T were unpleasant environments for the workers who toiled in them. Helping to assemble a Ford vehicle could be mind-meltingly boring, requiring staying in a single spot for an entire workday while performing the same individual motion time and again. Consequently, employee retention rates at Ford plants were abysmal.

What to do? Ford opted to pay his workers better. *Much* better. Pushed by his fellow executive James J. Couzens, Ford Motor Company revealed on January 5, 1914, that it would "inaugurate the greatest revolution in the matter of rewards for its workers ever known to the industrial world." This would include profit sharing, reducing the number of work hours from nine to eight per day and the biggest of the steps: doubling the prevailing wage to $5 a day. Controversy ensued. The *Wall Street Journal* called Ford a criminal, and *New York Times* publisher Adolph S. Ochs marveled, "He's crazy, isn't he?"

Despite the criticism, the new wage worked on a number of levels. Job retention increased. Profit sharing provided Ford employees with an incentive to work efficiently and well. Demand grew for Ford's Model T vehicles because now his own workers were better able to afford them, which in turn helped promote the convention of at least one car per family.

The incredible appetite for the cheap vehicles would transform the American landscape. Before the Model T, the existing oil business focused on the production of kerosene for home and urban illumination. Gasoline was considered such a useless by-product that refineries had pumped it straight into rivers rather than bothering to sell it. Then, just as Edison's lightbulbs curtailed demand for kerosene and thus for oil, Henry Ford's gas-powered vehicles kept the refineries in business. With their fuel tanks topped up at increasingly numerous filling stations, Ford's Model T's needed somewhere to drive, and urban planners paved over vast swathes of the American landscape, which led over time to suburban development, the daily commute, rush hour and, eventually, the creation of one of world history's biggest engineering projects, the American interstate highway system. Thanks to it all, vehicle ownership became an important component of membership in the American middle class. You hadn't really made it in America unless you owned your own car—no matter how wasteful and irrational that was.

The DNA of the machine created by men like Benz, Daimler, the Duryea Brothers, Henry Ford and later innovators like GM's Charles Kettering was extraordinarily powerful because it offered freedom on the cheap: You could go where you want to, when you want to, all at reasonable cost. I marvel at the parallels between the era that first set the stage for the auto industry, say, 1885 to 1925, and today. The dawn of the auto industry was an inflection point that saw nascent technologies shaking themselves out against a backdrop of cheap oil and the rapid construction of well-built roads. Today, similarly nascent innovations—autonomous technology, alternative-propulsion methods and the ride-sharing business model of transportation—will have a similarly transformative effect on the way people get around.

When, exactly, did the automobile become synonymous with the idea of America? The Second World War, when the car factories converted themselves to build planes and tanks for the American war effort? The fifties, as our nation explored the new interstate highway system in hot rods and fin-tail sedans? Or the sixties, when middle-class families traded downtown urban areas for the two-car garage suburbs designed to be more hospitable to automobiles than human beings? Each transition integrated automobiles more tightly into the national culture. "What was good for our country was good for General Motors," said GM CEO Charles Wilson during his confirmation hearings to become President Eisenhower's defense secretary, and the press flipped the statement around to say, "What's good for General Motors is good for the country." Probably it was true both ways; at the time, GM was the nation's largest employer, with a bigger workforce than the populations of Delaware and Nevada combined. But it was more than that. What's good for *Detroit* was good for America, it was thought. Sometimes it seemed like Detroit *was* America, and America was Detroit.

To understand Detroit, you have to understand the characters who populated it. The city and the industry were dominated by car guys. People who loved to drive, who loved cars and horsepower and engines and grease, who loved nothing more than the rumble of an internal-combustion engine, the force pushing you back in your seat when you pressed against the accelerator. Car guys were happiest with lungfuls of automotive exhaust, their hands blackened with engine grease. Some were executives or assembly line workers. Others were just hobbyists who considered deep knowledge of Detroit's product lineup integral to their identity. Some drove around with bumper stickers slapped onto their trucks or hot rods, the more offensive the better.

"The problem with all these people," reported one observer in David Halberstam's history of the automotive industry, *The Reckoning*, "is that if you opened the tops of their heads, instead of brains you'd find carburetors." These men—and most were men—dreamed constantly of ways to cram more horsepower under those steel hoods, to gloss the car's skin with more chrome, to provide more comfort and control to the car's driver, and if that meant a little worse vehicle mileage, well, so long as gas was cheap and the roads were clear, what did it matter? "The intelligentsia of America, much given to driving small, fuel-efficient, rather cramped foreign cars, often mocked Detroit for the grossness and gaudiness of its product," noted Halberstam. "To many liberal intellectuals Detroit symbolized all that was excessive in the materialism of American life . . . None of this carping bothered Detroit."

The prototypical car guy in my day was a blustering, silver-haired executive named Bob Lutz, a cigar-chomping helicopter pilot who ran GM's product development for a time and, while he did, never failed to be surrounded at industry events by a clutch of fawning media. Find an extra $500 million in the budget and a car guy like Lutz would use the windfall to build a sixteen-cylinder Cadillac with extravagant horsepower. On their own, car guys would never use that money to decrease auto emissions or increase the fuel efficiency of an engine. And they certainly would never think to devote that money to funding alternative-propulsion technologies.

Detroit featured other stock characters. There were, for example, the bean counters, the best known of which, likely, was Robert McNamara, the defense secretary who expanded the United States' role in the Vietnam War, and who was previously a president of Ford Motor Company, the first who wasn't a member of the Ford family. Like most bean counters, McNamara was a finance whiz who cared little for cars as anything but a way to make money. The car guys disliked the bean counters. But the people the car guys *really* despised were the reformers.

The reformers were men and women who wished to rein in the excesses of the auto industry. Who hoped to make the automobiles less gas-guzzling, less exhaust-spouting. Who wanted to halt the paving over of farmland, the bulldozing of neighborhoods for highways, and who hoped to pass laws that would require automobiles to become less deadly, so that fewer people would die in car crashes. That, at least, was the intent of one of the most seminal of reformers, Ralph Nader, a Washington, D.C.–by-way-of-Connecticut lawyer whose landmark 1965 book, *Unsafe at Any Speed*, called on the automotive industry to make cars safer, and became an unlikely bestseller. The book criticized poor brake performance, rigid steering wheel columns and nonexistent crash protection for making Detroit's vehicles more dangerous than they had to be. Nader's book triggered the development of the National Highway Traffic Safety Administration, which aimed not only to reduce fatalities in automobile crashes, but to prevent crashes in the first place. At the time Nader's book was published, about five people died for every 100 million miles driven, according to NHTSA. Today, that figure is about one per 100 million miles. Much of that difference stems from Nader's advocacy. For his criticism of the auto industry, Nader would become so reviled in Detroit that GM hired private investigators to tail him in the hopes of discovering information that might discredit Nader to the public. In the decades since, Nader has not gained any more friends in Detroit. Yet the effects of his advocacy for vehicle safety cannot be disputed. In the fifty-plus years since *Unsafe at Any Speed* was first published, Nader has saved millions of lives in the U.S. alone and for that reason was accepted into the Automotive Hall of Fame in 2016.

There were a few auto-industry reformers prior to Nader, but many more followed him. The reformers tended to take the current situation, extrapolate it forward and then point out how the existing trends were untenable, leading to global warming, energy crises or other misfortunes. Another common attribute among auto reformers was that they tended to exist outside of the auto industry, and in opposition to it. (And tended to neglect the beneficial effects of the automobile industry, such as the freedom the sector's products provided to users.)

My place on Rick Wagoner's automotive strategy board made me an insider in Detroit. I was one of the top people in the industry—and yet I, too, existed in opposition to Detroit's most unsustainable tendencies. I straddled both worlds—on the one side, perpetually working to wean the industry from oil and bloated vehicles and toward smaller "city" cars and electric propulsion technologies like fuel cells and battery-electric vehicles; on the other, sharing conference rooms with ultimate car guys like Bob Lutz. Often, it seemed like I had more in common with Nader than Lutz—and my own personal story does a lot to explain how my peculiar role came about.

I grew up in the Detroit area, in a Michigan dominated by car culture. Although my geography and my background should have made me live and breathe cars, I never felt like a car guy. My dad and his uncle owned a twenty-four-hour diner on Saginaw Street in the Detroit suburb of Pontiac—which provided the since-defunct GM brand with its name. The diner catered to three shifts of autoworkers from the nearby Pontiac Motor plants. From the time I turned eleven in 1962, I bussed dishes, waited tables and prepared food along with my brother, father and great-uncle, on weekends during the school year and five days a week in the summer. My shift tended to go from 5:00 A.M. to 1:00 P.M. Sunday mornings were always the most interesting. My brother and I would show up for work, and the first couple of hours would be these guys who had been up all night, drinking and gambling at Pontiac's Roosevelt Hotel. Then came the night-shift autoworkers. I was not then in the industry, but I felt closely connected with it because I was serving these hardworking men and women.

My other defining life experience was baseball. My father coached the local American Legion baseball team for boys sixteen to eighteen years old. Because we worked the morning shift, the afternoons allowed us to hit the diamond for practice. My older brother, Jim, was a good baseball player. I was not. But I loved the game, so my dad let me be the third-base coach, manage the equipment and keep score. I was his right-hand man in running the team's logistics. We ended up winning the Michigan state title twice, first in 1969 and then again six years later, in 1975. Kirk Gibson was on that second team to win the state title, and Kirk's dad was my dad's assistant coach. Gibson, of course, would go on to win the World Series in 1984 with the Detroit Tigers, and again in 1988 with the Los Angeles Dodgers.

During my teens, many of my friends spent their Friday nights cruising Detroit's Woodward Avenue in Pontiac GTOs or Ford Mustangs. I did my share of tinkering with engine parts, mostly thanks to a neighborhood friend's dad, a chief mechanic at an auto dealership, who would bring them home. I just loved to take the devices apart and then put them back together. But I could never see the sense in fiddling with the engines to make them accelerate more quickly or boost their top speeds. Making those hot-rod engines more powerful just seemed to put their drivers on a faster road toward calamity. Bob Lutz often joked that "it's not the speed that kills you, it's the sudden stop." Well, I felt it was both.

I graduated from high school in 1969, and at that point my older brother was already going to Eastern Michigan University and my older sister was going to Central Michigan. We weren't poor by any means. The restaurant gave us a solidly middle-class existence. But with two kids already in college, my folks didn't have much money to pay for a third college education. So I chose General Motors Institute in Flint, not because I had a burning desire to work in the auto industry, but because it operated on the co-op model. Students there attended classes for a month and a half, then spent the next month and a half working for General Motors for a good salary. GMI, which is now called Kettering University, gave me a chance to graduate from college with money in the bank. While there, I fell in love with math. I loved the problem solving that GMI taught, the beauty of the engineering thought processes: how to frame problems, and how to solve problems with analytics and data. Plus, the unconventional co-op schedule represented an efficient way to gain real-world experience while still attending school. I ended up doing a five-year program in only four years and finished second in my class in 1973.

It was a strange time to graduate from a co-op school that set you up to work in the world's largest automaker. My military draft lottery number was 283, which meant I was unlikely to be called to fight in Vietnam, but the war continued to weigh heavy on me, as it did for many others in my generation. The city of Detroit was struggling in the wake of the '67 riots. The ongoing Watergate scandal had triggered an overall suspicion of major American institutions. That fall, the Arab oil embargo would push oil prices up four times, from $3 a barrel to $12. Stricter emission regulations were problematic for Detroit, as were the lingering effects of Nader's book. Amid all that, the auto industry wasn't exactly the most popular choice of career.

While many of my fellow top GMI graduates would go on to Harvard, where GM paid for their MBAs, I was thinking about questions other than how to make money selling cars. I wanted to explore the transportation system and why so many people chose the automobile to get around. On a scholarship from General Motors, I went to the University of Michigan to study public policy. My program was tailored for engineers with an appetite to learn about economics and politics. Once I received my master's in 1975, I joined GM research and development full-time as an engineer. But I quickly realized that if you really wanted to have freedom to pursue your priorities at GM, you needed to have your PhD. So I headed to the University of California at Berkeley. My research, I decided, would be in transportation systems from engineering, economics and policy standpoints. Not only how people get around—but how we could make that system more efficient.

At that time, I was driving a Volkswagen Beetle with flowers painted on the sides. To get out to Berkeley, I bought a Chevy van that I customized for camping. I headed west in the fall of 1975— bearded, shaggy-haired, not exactly a full-on hippy but pretty close. I took an educational leave from GM and remained employed there in the summers. They generously covered my travel expenses to and from California and rather than go by air, I cashed in the tickets and drove, allowing me to see the country: Rocky Mountain National Park, Yellowstone, the Grand Tetons, Yosemite and the Grand Canyon—you name it, I've camped there. Once I finished at Berkeley in 1978, MIT offered me a position as an assistant professor. But GM made me a job offer, too, and I felt like I had a duty to go to work for the company that had employed me since I finished high school. So I filed my dissertation within an hour of the deadline, hopped into my van and drove back to Michigan. I began a position at GM Research Laboratories in July 1978, telling myself I would stay for just a couple of years. I ended up staying for three decades.

Working for what was then the world's largest automaker gave me a front-row seat to observe the auto industry's irrationality. Within GM I became known as a researcher who could creatively use math to solve tricky problems. In 1988, the guy who was in charge of all the technical workers at GM, who would later go on to run Chrysler, Bob Eaton, called me into his office and asked me if I'd thought much about what I wanted to do with my career. I said, "Well, I educated myself to do research—and that's what I really enjoy doing." Eaton said, "Well, we have some other ideas."

Eaton put me in charge of resource planning and production control for the Buick, Oldsmobile and Cadillac Group, working under a brusque but kindhearted executive named Don Hackworth, who would become one of my best friends within the company.

I saw how unproductive we were in our approach to product development, and how poor our quality was relative to the competition. After WWII the auto industry grew very quickly. Profits were so high that the inefficiencies and lack of quality were masked by the huge societal thirst for cars. Then came the oil crises of '73 and '79, and a growing awareness of the impact of automobiles on air pollution. The bustling economy of the eighties allowed us to battle overseas imports without addressing the waste and inefficiency. But when consumer demand slowed in the early nineties recession, Detroit was in trouble. General Motors in particular was hemorrhaging money. From the calculations I'd conducted for Hackworth, I knew we had twice as many workers as we needed—a tricky situation when our agreement with the United Auto Workers guaranteed employment. Hackworth put me on a secret committee that decided which plants to close. The process was so controversial we made only one handwritten list of the candidates. It was on a single sheet of paper that I kept with me at all times. One day the news broke that we were planning to close a plant in Wilmington, Delaware. The media portrayed us as an unfeeling corporation robbing honest workers of their livelihoods. The following morning Hackworth seemed really down when I saw him. When I asked if the press coverage was bothering him, he said, "No, Larry, it's not that. Last night, after the news broke, one of our hourly employees in Wilmington committed suicide." It affected me, too. How could it not? But GM was in a tricky situation. If we didn't cut costs by closing factories, we were headed for bankruptcy.

In fact, I thought we *should* have gone bankrupt in the early nineties slowdown. The deals we'd signed with the United Auto Workers meant we were basically a health care company that just happened to make cars. Bankruptcy would have let us negotiate new labor agreements and shed the legacy costs required by our health care and pension obligations. Instead, we limped through the crisis. I spent the mid-nineties working as head of planning for Rick Wagoner, who at the time was president of GM North America, to try to make GM more efficient. How should we develop our products? How should we improve our quality processes? Whatever it was, Rick would throw these challenges at me. Some of the things I tackled seem a bit insane in retrospect. For example, we had an annual process to create a ten-year plan that took fifty-three weeks to complete. (I turned this into a five-year plan, because planning six and even ten years out seemed like a waste of time. I also developed a process that updated the plan daily if necessary, rather than yearly.)

During the same period, I encountered a personal crisis that transformed my thinking about the power of technology. Years before, when I was twenty, I'd lost the hearing in my right ear. It was a bit unnerving, and I saw a doctor about it, but no one could figure out the reason it happened. I adjusted soon enough to hearing out of only one ear, and years passed without me thinking much about it. Then one Thursday evening in 1993, after watching an episode of *Seinfeld*, I went to bed, only to wake up later that night absolutely deaf. My wife, CeCe, drove me to the hospital and once again, they couldn't figure out what the cause was. She called Hackworth later on Friday and told him the news. Don told her he'd see me in the office that Monday, which seemed a little heartless. I mean, I was pretty freaked out by the possibility that the last thing I'd ever hear was the bickering of Jerry, George, Kramer and Elaine. But as it turned out, Don's response was exactly what I needed. It meant that I couldn't dwell on going deaf, or get depressed about it. Instead, I immediately focused on dealing with the disability and getting on with my life. When I showed up for work on Monday morning, I saw that Hackworth had hustled over the weekend to get a stenographer to attend meetings with me and write out what was being said, so I could understand what was going on. He'd prepared my secretary for the change and informed his staff what to expect. I also took classes to learn how to read lips. It helped me communicate with everyone except a colleague of mine, Tom Brady, who ran all of GM's stamping plants and had a mustache so bushy I couldn't see his mouth. There was one lunch meeting where Tom was chewing a piece of gum as well as a toothpick and I jokingly told him that it was hard enough to read his lips due to his mustache—but it was impossible with the gum and toothpick. He laughed, then disposed of it all in the trash—and then, the next day, he showed up at work with a clean upper lip. He had shaved off his mustache just so I could understand him better.

What made me the optimistic technologist that I am today was the cochlear implant that I got twelve months after I'd gone deaf. I believe I was among the first thousand people in the world to receive one. Activating the device was overwhelming. The fidelity wasn't great. The sound quality was similar to what you might get from a poorly tuned AM station. And after the silence of the previous months, the world seemed a cacophony of unrecognizable noise. Five minutes in I left my wife and the audiologist to go and use the bathroom. The first sound I really recognized was the noise created when I stood in front of the toilet. "This is going to work," I told my wife when I returned to the room. The following month, in June 1994, the first spoken words I understood were part of a broadcast on my car radio about O. J. Simpson on the run in his white Bronco. Having my hearing returned by this remarkable technology encouraged me to perceive the way science could solve mankind's ills, including, just possibly, the irrationality that orbited the automobile. It reinforced my belief in the potential of technology to improve people's lives. And, living deaf for a year helped me better understand why disabled people strive so hard to be independent. Like everyone, they value their freedom—their autonomy.

In 1998, after I'd been with GM for twenty years, the company's then-president and chief operating officer, Rick Wagoner, placed me in charge of research and development, directing an annual budget of about $700 million. The promotion came as a surprise. I really liked my current job, handling planning for North America, and Wagoner told me I could keep that responsibility. My new title was corporate vice president of research, development and planning. I was thrilled.

The promotion required me to map the company's distant future. And because General Motors at that point was the world's biggest automobile manufacturer, the job provided me with an opportunity to significantly influence the direction of the industry.

To understand what I was getting into with my R&D promotion, you have to realize that I came in just as GM was about to get its nose bloodied by the shutdown of the EV1 program, which had created this groundbreaking product, the first modern mass-produced electric vehicle to be released by a major automobile manufacturer. A lot has been written about the EV1, and there's even a fairly witty documentary about it, *Who Killed the Electric Car?*, so I'll avoid repeating what's already out there.

Here's my take—and how it set the stage for my term as GM's R&D chief. The EV1 was a response to changes in California's emissions policy. Early in the nineties, California state legislators were going to require that a certain proportion of vehicles sold in the state have zero-emissions. So GM's CEO at the time, Jack Smith, cut a deal with California: If we made our best shot at a pure battery-electric vehicle, could we delay the zero-emission-vehicle requirement, to give us a better chance to prepare for it as an industry?

The state of California agreed. GM's Advanced Technology Vehicles group came up with a remarkable two-seat coupe with a lead-acid battery pack that could go 50 miles on an 8-hour charge. Not impressive numbers for today; the Chevy Bolt can go 238 miles on a 9.5-hour charge; Tesla's Model S has a range of 341 miles on about 6 hours if you use Tesla's proprietary wall connectors and employ dual chargers. But for its time the EV1 was pretty amazing. We began leasing it in 1996, two years before I led GM R&D, and the people who drove it loved it. *I* loved it. It was incredibly aerodynamic, with a drag coefficient of just 0.14, when a good conventional gas-powered car had a drag coefficient of 0.3, helping it to feel smooth and fast. Accessories like headlights and the heating and cooling had to be ultra-efficient, to leave as much energy as possible to power the movement of the vehicle. With a traditional gas-powered car, there's a short pause after you depress the accelerator, before it actually moves forward. Electric vehicles don't have that delay. The EV1 was many people's first exposure to the remarkable responsiveness of electric vehicles. You simply nudged the accelerator pedal and the thing shot down the road.

The problem was that it was ungodly expensive to make. We spent more than a billion dollars designing, engineering, tooling and making just over one thousand vehicles that customers leased from 1996 to 1999. That's a cost of nearly a million dollars per vehicle. Just a few years past our near bankruptcy in '91 and '92, many people in the company considered the EV1 irresponsibly profligate. They felt we needed to reserve the budget for things that made us money *immediately*, or at least in the next year or two. We couldn't keep a car in production when making it was losing us hundreds of millions of dollars a year. Not when demand for a two-seat battery-electric vehicle with a fifty-mile range was so small. Consequently, Wagoner decided to stop investing in the EV1 program. In the same conversation when he promoted me to VP of R&D, he asked me to re-deploy the Advanced Technology Vehicle group to other programs.

Wagoner would later call the curtailing of the EV1 program one of the bigger mistakes he made at GM. Rick remains a good friend of mine. I still see him regularly, and I respect him enormously. I also agree with his assessment. When we halted EV1, General Motors was likely five years ahead of everyone else on battery-electric vehicles. We had two generations of improved batteries in the pipeline, including nickel metal hydride for production development and, a little further off, lithium-ion in technology development. (Nickel metal hydride featured half the weight and twice the energy-storage density of lead-acid batteries.)

Stopping the EV1 program brought up a new difficulty. Lead-acid batteries had never been used to energize an electric-propulsion motor in a modern consumer automobile. We were certain the batteries were safe, but just to be sure, we leased the vehicles rather than selling them outright. And when we halted the program, it was prudent from a safety and liability standpoint to take the vehicles back as the leases expired. But because a lot of high-profile people in California were big fans of these cars—including Mel Gibson, Tom Hanks and Ed Begley, Jr.—there was a huge public outcry when we asked for them back. Even worse, there was the issue of what to do with the one thousand EV1s once we'd repossessed them. So someone outside GM R&D made the decision to crush the vehicles, to recycle them, and someone else got the crushed vehicles on camera—and the footage made it into the documentary I mentioned. Which led people to believe that GM intentionally conspired to kill the electric car. It was a public relations disaster. We spent a great deal of money to take leadership in this technology, and then thanks to the way we shut down the program, we created an image for ourselves that was the exact opposite of a company interested in environmental stewardship.

In hindsight, we should have pivoted the EV1 program into a hybrid vehicle. A few years later, with the hybrid Prius, Toyota is said to have accepted that the first round of vehicles were going to lose money—and then they improved on that first generation. Had we accepted that profitability wouldn't come until several vehicle generations down the road, had we accepted that this was something worth doing for the long term— then we could have engineered a hybrid gas-electric powertrain, put it in the EV1 platform, added a backseat and been on the U.S. market years before the Prius. Meanwhile, Toyota got a few more generations' worth of learning ahead of us, *and* they became known as the industry's green automaker.

We blew it with the EV1. To be frank, we blew it because of the short-term pressure of rewarding shareholders with appropriate returns, the health care and pension costs hamstringing us in the early nineties and the need to do a whole lot of spending on our fundamental business to get back in the game. Tesla's first car, the Roadster, debuted fifteen years after EV1 and wasn't nearly as innovative as the EV1 had been in its era. Electric-car technology would be a lot further along than it is today if GM had kept that program going.

GM had historically been the industry's technological leader. In the aftermath of the EV1, in 1999, Rick thought we needed a way to take back that role. Add to that the fact that the turn of the millennium was coming up. The automobile was now more than a hundred years old, and General Motors itself would in 2008 mark its centennial anniversary. We'd weathered the early-nineties' setbacks. The profits that obscured our inefficiencies were back due to strong SUV and pickup truck sales. Now Rick was thinking about how to regain GM's historical position as the company that defined the rest of the industry's direction.

Over lunch one day, Rick and I got to talking about how little automobiles had changed since the industry's birth. Gas-fueled, run by an internal-combustion engine, rolling on four rubber tires, with the passengers protected by a windshield and four doors: That was the paradigm used by the Model T, and it was the same paradigm we were using today. Rick wondered if there was a way to get rid of the negatives of auto ownership and yet still preserve the absolute essence of what auto mobility stood for: spontaneity, responsiveness, freedom and personal use. The liberty that comes with having a car. How could we give our customers all the good things without any of the bad things?

Rick asked me: What's the car of the next hundred years going to look like? If I were going to use current technology to create the automobile—if the automobile were being invented *today*—then what form would it take?

Thus began the most exciting research initiative of my career to date—and the development of a vehicle that represented the first step in the mobility revolution. It was my opportunity to re-envision the way Americans move around and interact. After years of privately bewailing the industry's inefficiency and waste, I had an opportunity to put GM's money where my mouth was. It was a once-in-a-career chance for a mobility guy to take on the car guys and bean counters.

Chapter Five

EPIPHANIES

It is not an optical illusion. It just looks that way.

——STEVEN WRIGHT

The result of our reinvention of the automobile debuted four months after 9/11, at the 2002 North American International Auto Show in Detroit. For GM, the event was the industry's most significant. It's where the biggest announcements happen before the biggest and most important crowds that are nursing the biggest hangovers from the biggest no-holds-barred parties. There's media, there's industry executives and there are members of the regular public. And I was there about to introduce another superlative: one of the strangest concept cars the auto industry had ever seen.

I stood behind the curtains as Rick Wagoner walked out onstage before a jam-packed crowd. Next to me was the result of two years of work and countless more years of research. It had four wheels, like most vehicles, and that was the sum total of its resemblance to a conventional automobile. In between the four wheels was a flat, sleek slab, maybe six inches thick. The tires were narrower than most automobile tires, and slightly taller, at about two feet high. From ground level, the chassis bore a resemblance to a longboard skateboard—except it was big enough to support an SUV.

"Today," said my boss, on the other side of the curtain, "we're making what we believe is one of the most significant concept-vehicle announcements that GM has made—at this show, or any show."

Most people on the curtain's other side were expecting us to roll out some new version of the Malibu, say, or an even more futuristic version of the Corvette. I could feel them perking up. What was the CEO of the world's largest auto company up to, making a pronouncement like that? Now that Rick had made the promise, he had to deliver.

I reached out and ran a hand across the chassis's smooth surface. "In a minute," Rick said, "we will introduce a revolutionary concept—so revolutionary that we believe it's no stretch to say it could literally *reinvent* the automobile."

Rick got a few paragraphs further through the speech I'd helped write. I double-checked the batteries of my cochlear implant and made sure my pants zipper was up. And then I heard him say, "It's my pleasure to present . . . Autonomy."

The chassis rolled out past the curtains bathed in a purple light that reflected nicely over the gray composite housing.

Silence. And then there was this kind of collective gasp. Curious— that's what the audience were. They were waiting. We'd rolled out this strange car-size skateboard, and they were intrigued enough to indulge us, to give us a chance to explain what it was about. That's the moment I strode onstage.

"We called our concept Autonomy," I said, and paused. My throat was dry. I cleared it. "Because freedom is what automobiles are all about.

"I'm talking about the freedom to go where you want to, when you want to—and with the people and things you desire to take along . . . To reinvent the automobile around today's technology, we started with a clean slate," I said. "Our opportunity was to create an all-new, logical and exciting design from scratch.

"We landed on an innovative vehicle architecture with a six-inch-thick skateboard-like chassis," I said, as the chassis rotated horizontally on a platform a few feet away. "It has electric motors in all four wheels and the fuel-cell stack, hydrogen storage system, controls and heat exchangers are embedded within.

Then I said what Autonomy *didn't* have. "There is no internal-combustion engine. No transmission. No drivetrain, no axles, no exhaust system. No radiator, and no mechanical steering, braking and accelerating linkages." Then I paused. "In fact, the only things moving other than electrons, protons, water and air—are the wheels and suspension!"

That was the remarkable thing about this concept. It was designed around a hydrogen fuel cell. Autonomy was among the first vehicle designs to fully take advantage of the potential offered by alternative-propulsion systems. The new technology liberated engineers and designers who had worked on the prototype, including GM's resident fuel-cell expert, scientist Byron McCormick, as well as Autonomy's program manager, Chris Borroni-Bird, and GM's design team. The chassis, which we referred to as the "skateboard" it resembled, housed all the mechanical and powertrain parts. We'd managed to fit all the components inside the slab. Nothing mechanical had to sit on the chassis because we'd digitized and miniaturized most every part required by the vehicle—and crammed it into that six-inch sandwich. It was there that a tank of pure hydrogen gas energized a fuel cell that melded together hydrogen and oxygen atoms, resulting in electric energy and water. It was in there that four electric motors, one at each wheel, accelerated the vehicle forward.

Moments later, out came another version of the Autonomy concept. This one looked more like a recognizable vehicle. On the chassis we'd set a sleek, gray auto body that looked a bit like a Formula 1 race car with air intake scoops at the front. Then we demonstrated how the lightweight vehicle body could be lifted from the chassis, to be replaced moments later by another body. So in this way, the owners of the chassis might opt to use a sporty coupe version when going out to a dinner date—and then, the next day, replace it with a larger-capacity, SUV-type body to transport a load of kids to soccer practice.

The hydrogen-fuel-cell technology had been invented in 1842 by a British lawyer and amateur scientist named Sir William Grove. GM had a long history of working on vehicles related to fuel-cell technology, having created the first earthbound hydrogen-fuel-cell vehicle, the 1966 Electrovan. The problem was, the Electrovan employed *alkaline* fuel cells, which tended to be so bulky that the van barely accommodated any passengers. McCormick knew about a less bulky alternative to alkaline fuel cells that had been invented by GE scientists for the Gemini and Apollo space missions.

Onstage, at the Detroit auto show, I described the advantages that fuel cells offered. Not just that the only by-products of the vehicle's workings were pure water and heat—no carbon dioxide, no nitrogen oxide, either—it created no fumes at all and you could drink the water that dripped from its tailpipe. And the skateboard approach to vehicle construction could allow GM to streamline its manufacturing. We could get by with only a few versions of the Autonomy skateboard. Perhaps one would exist for two-person vehicles. We could have another for four-to-six-passenger vehicles, and a heavy-duty version for pickup trucks and the burliest of SUVs. The passenger compartments that were lowered on top of the skateboards could feature potentially infinite variety.

"Today, one hundred years after the automobile's invention," I said, "only twelve percent of the people in the world enjoy the freedom benefits of personal transportation . . . Autonomy offers the promise of extending this freedom to even more people."

The crowd seemed to love the idea of the Autonomy concept. The press did, too. Coverage in the *New York Times* and *The Economist* referred to "the car of the future" that could "represent the most dramatic change in car technology since Carl Benz first chugged out of his garage in 1885." At the same auto show, the Bush administration revealed plans for a program called FreedomCAR, designed to reduce America's dependence on foreign oil by investing in fuel-cell technology. The Autonomy project gained traction the following year when President George W. Bush made hydrogen-powered cars a key point in his 2003 "State of the Union" address. "Our scientists and engineers will overcome obstacles to taking these cars from laboratory to showroom, so that the first car driven by a child born today could be powered by hydrogen, and pollution-free," the president said. He even posed for photographs before the Autonomy prototype.

The New Yorker's environmental writer Elizabeth Kolbert was gushing in her praise for fuel cells in cars. "Outside of science fiction, the hydrogen car is probably the most radical reinvention of the automobile that has ever been imagined," she wrote. "It can even serve as a source of electricity: at night, a hydrogen-car owner can use his vehicle to light his home."

As Kolbert wrote, the Autonomy prototype was my attempt at a "leapfrog" technology, which "does not merely improve on something that exists but vaults over all possible improvements to achieve something totally new." I liked that term, "leapfrog," and would think of it often as I oversaw similarly radical projects in the future.

After the 2002 Detroit auto show I asked Autonomy's project manager, Chris Borroni-Bird, to work with Byron McCormick's team to create a more practical vehicle. The result was the GM Hy-wire, the first drivable automobile prototype to combine fuel-cell technology with by-wire controls, that is, controls that used electric actuators and software to manage steering, braking and speed. We introduced it in September 2003. Again we used a skateboard-design approach, setting the passenger compartment on a chassis that featured all the major powertrain components. That meant the passenger compartment had very few design restrictions. The windshield was enormous. Video cameras provided a 360-degree view rather than mirrors about a decade before that became commonplace. The steering mechanism was more like the controller for a driving arcade game, with two handles at what would have been a clock's nine and three positions, one of which rotated to control the speed. Hy-wire had so much legroom the vehicle could have comfortably accommodated four pro basketball players. And no hump in the middle of the vehicle at all, so sitting inside felt like you were relaxing in a well-appointed room that just happened to feature bucket seats.

By this point I was getting very enthusiastic about fuel-cell automobiles. It seemed like every month Byron McCormick would invite me down to his lab to demonstrate some other new innovation. The lab had been able to increase the power density of the fuel-cell stack by seven times in just a couple of years, which meant the stack could be smaller while still providing more power. At first fuel cells were fussy at low temperatures—a real problem in a place like Michigan. But McCormick was improving the range of temperatures that fuel cells would provide power. And the cost of the technology also was going down rapidly.

My advocacy of fuel cells, and alternative-propulsion systems overall, made me something of a curiosity in Detroit. Car guy Bob Lutz famously maintained that global warming was "a total crock of shit." Meanwhile, I was the rare auto executive who acknowledged climate change. "Burns is generally credited throughout the auto industry as the champion behind GM's goal to realise sustainable mobility with compelling and affordable vehicles," ran one 2003 profile of me. "If it's true," I told the *New York Times* in 2002, "the ramifications [of climate change] are so significant that it's irresponsible not to be addressing it."

My outspokenness drew some criticism from inside GM, whose best-selling products ran toward big and not particularly fuel-efficient SUVs. The media loved the controversy. My reputation as Detroit's biggest advocate for alternative-propulsion technologies is probably why *The Economist* invited me to speak at their 2004 oil and gas roundtable, which attracted senior figures from the oil industry in Houston.

I felt a bit like I was headed into the belly of the beast. So I put together a presentation that didn't pull any punches, describing the necessity of a wholesale reinvention of the auto industry, a theme I would revisit many times over the ensuing years. "General Motors believes there are many compelling reasons to move as quickly as possible to a personal mobility future energized by hydrogen and powered by fuel cells," I began. "Since only twelve percent of the world's population currently own automobiles, there is significant opportunity for growth in the future," I noted. "GM would like to make it possible for the remaining eighty-eight percent of the people who don't own vehicles to be able to enjoy the benefits of ownership—to be able to go where they want, when they want, carrying what they need." And to do that without making a major contribution to global warming, I told the group of oilmen before me, we needed sustainable vehicles. "With these trends," I said, "it's critically important to develop new energy sources for transportation."

As I was speaking at the Houston oil conference, the team of McCormick and Borroni-Bird were busy working on a new vehicle, called the Sequel—a fuel-cell prototype designed to travel on public roads. This, too, was built over a skateboard, but in most other respects it looked more like a conventional car. Visualize a supersize version of the Chevrolet Bolt and that's pretty much the Sequel, which automotive journalists called a "sport wagon." It was a combination of two alternative-propulsion methods. It had a lithium-ion battery pack as well as a fuel cell. The propulsion was electric drive: A single electric motor drove the front wheels, while extra power for acceleration came from smaller motors that powered each of the rear wheels. When the Sequel braked, the inertia in the rotating wheels generated electricity that was then stored in the battery to power future acceleration—a fairly common technique today.

Those were tumultuous years for the world. From the consequences of 9/11 to global warming, wars in Afghanistan and Iraq, and oil spills in fragile saltwater ecosystems—all of it was tied in some way or another to the gasoline we pumped into our vehicles. Electric vehicles seemed the answer to many of the problems the auto industry faced. Yet some of my peers in GM criticized the money I was spending on alternative-propulsion systems. I spent years feeling like we were just a board meeting away from having our budget not just cut, but eliminated. Meanwhile, we were spending much more as a company to make the internal-combustion engine more fuel efficient and with fewer emissions. All this work was pretty costly—GM's advanced-technology budget in 2005 was nearly three-quarters of a billion dollars. But it also was leading to some serious epiphanies. Take the skateboard vehicle architecture underlying Autonomy, Hy-wire and Sequel. That's very similar to how Tesla builds its vehicles today.

Another epiphany that arose from our advanced-development research was more troubling for General Motors, and the auto industry in general—while being great for the regular, everyday people who bought and used the cars we made.

Byron McCormick was the one who first showed me the implications of this epiphany around 2005. At that point his team was developing something called the E-Flex Architecture, which was our attempt to create a single skateboard that could accommodate any species of electric vehicle, whether you're talking a battery-electric, a gas-electric hybrid or a fuel-cell vehicle. Once Byron's team had come up with a prototype, he asked me to come and see him. That was fine. I always had time for Byron.

But what was unusual was *where* he wanted to see me. He wanted to meet in the Vehicle Assessment Center, which was a massive warehouse about the size of five football fields where our engineers broke down into their component parts any vehicle we found interesting. The assessment center helped us understand everything, not only about how *our* vehicles were built and worked, but how our competitors' cars were built and worked as well. Let's say the BMW Mini Cooper had just come out. We would buy one from their dealer, drive it over to the assessment center and then our mechanics would disassemble it—take it *completely* apart, down to the last nut and bolt. And as the mechanics worked, they'd arrange the parts in an organized fashion around the skeleton of the vehicle. After one of these tear-downs, the vehicle looked like it had blown up in the world's most organized explosion.

When I visited Byron, he was set up in an area of the assessment center that had three bays. The first bay, he explained, featured a completely disassembled Chevy Malibu. I could see the front and rear bumpers, the seats and the four doors, as well as smaller, more mechanical parts, like a disassembled radiator and each individual piston. There are something like ten thousand parts in a Chevy Malibu, and I could see pretty much every one set out before me.

"Okay," I said. I knew the Malibu well. And I wondered what Byron was getting at.

The next vehicle he showed me was a disassembled second-generation Toyota Prius hybrid. This was an even more complex vehicle than the Malibu, in some respects, because it didn't just have a conventional internal-combustion gas engine in there—it also featured electric motors and the battery packs to power them. The pile of parts that could be assembled back into the Prius was even bigger than that of the Chevy Malibu.

"Great," I said to Byron. I mean, anyone who knew anything about the technology behind hybrid engines would know that the Prius would have more parts than the Malibu.

"But wait," Byron said. And then he brought me to the third bay. And I immediately saw his point. Just one look at the parts spread out before me was enough to explain why Byron had summoned me to the assessment center.

"Rick's got to see this," I said.

"I think so, too," Byron said.

So I set up a meeting with Rick Wagoner. I took Rick through the same process that Byron had taken me through. By the time we arrived at the third bay, Rick was nearly as terse as I was. "Great—it's an E-Flex Architecture disassembled," Rick said. "So what?"

Wagoner recognized a vehicle—which we called E-Flex—that was designed to accommodate numerous different powertrains, from battery electric to fuel cell to hybrid. The version of the E-Flex in the assessment center was powered by a hydrogen fuel cell and controlled with by-wire technology. Say a conventional gas-powered vehicle like the Chevy Malibu had ten thousand parts. The fuel-cell prototype had maybe one thousand, one-tenth the parts of the Malibu. McCormick was showing us that an alternative-propulsion vehicle could have an order of magnitude fewer parts than a conventional car. And the implications of that were enormous for General Motors, as well as pretty much every other automaker and supplier on the planet.

The auto industry is set up as a hierarchy. At the top are the companies whose name brands are on the cars everybody drives—GM and Ford, say, as well as companies like Fiat Chrysler Automobiles,

Volkswagen, Honda and Toyota. They design and assemble the cars. But these auto brands don't actually make all the parts of the cars they're assembling. Under them are auto suppliers, which have names like Robert Bosch GmbH, Denso, Delphi, Visteon, Continental AG and Magna International, whose size as corporations can rival those of the automakers themselves. And underneath them are smaller, regional suppliers who can supply the mega-suppliers with parts. It's a vast matrix of interconnected relationships and supply chains.

The brands are on top because they control the vehicle design and engineering specifications, which depend both on what the vehicle must do to excite customers and how the vehicle gets built. It's devilishly tricky to assemble something as mechanically complex as a modern automobile in a way that will ensure it will continue to run a decade after it's made. Conventional automobiles have thousands of moving parts because most of their controls are mechanical. Acceleration, steering, braking, transmissions—they all require many different parts to work. And each of these moving parts needs to be developed, product tested, tooled, sourced and engineered to be as reliable to work on its ten-thousandth use as it is on its first. That requires a lot of know-how.

Few other industries are able to manage the feat. The airline industry makes planes that have more parts, sure. Same with the shipbuilding industry. But they produce their enormous vessels in much smaller numbers. The automakers are the only ones to make such complex machines in such great numbers. It's a challenging task to tune a gas engine so that it feels responsive and powerful, while simultaneously meeting emissions regulations, fuel-efficiency standards and safety rules, or to assemble thousands of parts that fit together precisely. The point is, you need many, many engineers to create a well-designed car. It's *their* brains that design the cars in the first place. *Their* brains that lead the innovations. And *their* brains that differentiate a car that is fun to drive from a car that just feels like another bucket of bolts and parts.

What McCormick saw, and what we were now showing Rick,

was that these new electric vehicles were much simpler than conventional automobiles. With just one-tenth the parts, the electric vehicle was a lot easier to make. Not only did an electric vehicle have fewer parts, it also had far fewer *moving* parts than a conventional automobile. Take the gasoline out of the vehicle, and you didn't need a heavy engine block to contain any explosions happening within the cylinders. You didn't need an exhaust system with a muffler and a catalytic converter to clean the vehicle's emissions. You didn't need spark plugs or carburetors, or valves or fan belts to cool it. Nor did you require a fuel-injection system or an automatic transmission.

This E-Flex Architecture had an electric motor in the front of the car, which featured electromagnets, ball bearings and a spinning axle. At the back each wheel featured wheel-hub motors, which had similar components. In the middle there was a fuel cell stack and a hydrogen storage container and a heat exchanger, which had relatively few moving parts. In the same space, if this was a battery-electric vehicle, would be the batteries and the charging equipment. And there was something we called a "controller," which basically was a computer chip that had been programmed with software that made everything work together.

"This is all the parts it requires?" Rick asked.

"That's right," I said.

Rick gave a low whistle.

What Rick Wagoner got in that moment was the profoundly different scale of manufacturing an electric vehicle versus a gas-powered one.

Electric vehicles would require far fewer employees to assemble the vehicle. Like, one-tenth the number of employees. And because they were comparatively simple to make, they'd open up the manufacturing of America's vehicles to more brands. Detroit was about to get a lot less exclusive.

And not only that. Some of my early work with GM R&D, back when I was actually doing research, established that the biggest automobile cost driver, the real factor jacking up the sticker price, was the number of parts. Once we were mass-producing these alternative-propulsion vehicles in numbers that rivaled gas-powered cars, the electric vehicles would be much cheaper. And probably more reliable.

The other thing Rick grasped pretty much instantly was the crucial part of the expertise in alternative-propulsion vehicles. He looked at the exploded vehicle in front of him, at the parts scattered all around, and realized very few of them looked like the sorts of parts that he was accustomed to seeing in an automobile.

With *electric* vehicles, the crucial expertise was no longer going to be mechanical. Not with an electric engine. Not with drive-by-wire controls.

The crucial expertise in electric vehicles was software. So shifting to alternative propulsion would probably require a radically different General Motors. A radically *smaller* GM, which would use substantially fewer suppliers to sell substantially cheaper products. These projects would be engineered not by experts in the vagaries of the internal-combustion engine, but by programmers expert in the same ones and zeroes that ran your computer, your television and any number of computer-equipped devices around your home.

"What you're showing me," Rick said slowly, "spells the end to the integrated auto industry as we know it."

McCormick, Wagoner and I realized that technology was bringing us nearer to an important inflection point for Detroit. Something similar to what IBM faced in the eighties as its industry transitioned from minicomputers used and owned by businesses to personal computers used and owned by private individuals. During that transition, IBM outsourced its chip production and software coding to two companies: Intel and Microsoft. Not realizing that the real value in a computer came from the chips and the coding—and pushing IBM to the brink of irrelevance.

With electric vehicles, GM had to avoid outsourcing its valuable expertise to suppliers. A GM that continued with business as usual risked becoming little more than a packager. A tuner. A sales channel, which took other companies' valuable expertise and basically repackaged it into the vehicles bought by consumers. That moment at the assessment center was the first time I sensed a future that saw technology wreaking a transformative change on Detroit, and on the entire auto industry.

In 2007, though, alternative-propulsion technology needed to pass numerous milestones before it threatened the dominance in Detroit of internal-combustion engines. One of those milestones happened the morning of May 15. The sky featured ominously dark clouds in Rochester, New York, where I'd gathered a half-dozen auto industry journalists along with a handful of GM engineers to try to break a world record: the longest-ever drive in a fully functional automobile powered by a hydrogen fuel cell. To address the range anxiety that comes with any alternative-propulsion vehicle, I'd always told my team that we needed to be able to drive more than three hundred miles without a fill-up or recharge, roughly analogous to the distance some cars could travel on a tank of gas. (Although the more fuel-efficient vehicles today can travel much farther.)

It may have taken five years, but Byron McCormick, Chris Borroni-Bird and their teams had done it—they had engineered a fully capable vehicle, the Chevrolet Sequel prototype, that their calculations suggested could reach my benchmark with a single tank of hydrogen. We introduced the Sequel in 2005 at the North American International Auto Show. "While Autonomy and Hy-wire were concepts, Sequel demonstrates that our vision is now real—not yet affordable, but definitely doable," I told the crowd.

Now, in New York, we were going to put Byron's and Chris's calculations to the test by driving the three hundred miles from GM's Fuel Cell Activities Center in Honeoye Falls, just south of Rochester, to Tarrytown in Westchester County.

Before the hundred or so people gathered outside our building, I gave the same sort of speech I'd been giving each time we demonstrated the Sequel. "Our vision was to exchange the automobile's current DNA—the internal combustion engine, petroleum, and largely mechanical systems—for a new DNA . . . We truly believe that the transition from a mechanical vehicle to one that is almost entirely electrical is as momentous as the transition from horses to horsepower."

Eight and a half hours later we rolled into Tarrytown's Lyndhurst mansion, a voyage of 304 miles, with two pounds of hydrogen still in the Sequel's tank, enough for the vehicle to go an additional forty or fifty miles.

"I'd like to propose a toast—with water—which represents the only emissions that were created during our three-hundred-mile drive," I told the small crowd that had assembled to receive us.

Soon after the Autonomy debuted in 2002, I'd said that I expected GM to be making fuel-cell vehicles commercially available by 2010. I'd known back then it was a stretch goal—the sort of thing an executive like me has to say in order to motivate the researchers working on the technology. Had certain key things not happened—like the 2008 recession, and the Obama administration's switching of funding from fuel-cell development to battery-electric vehicles—we would have been a lot closer to reaching my goal. The closest that GM came under my watch were the one hundred hydrogen-powered Chevrolet Equinoxes that we scattered around the country in 2007 for Project Driveway, which studied the use of fuel-cell vehicles in the real world, driven by real people. Some fuel-cell vehicles were available in the U.S. for consumers within the time window I'd predicted. Honda began leasing its Clarity fuel-cell vehicle in the United States in 2008. Mercedes-Benz leased its F-Cell four-door sedans in 2010. Hyundai and Toyota also have their own vehicles available to consumers in certain areas where hydrogen filling stations are available.

Nevertheless, much of the learning we developed in those years could be applied to the development of other electric vehicles, like the Chevrolet Volt, a plug-in hybrid electric, the range of which can be extended with its on-board gasoline engine, and which was first sold in late 2010; or even more notably, the Chevrolet Bolt, the all-electric vehicle that GM released in late 2016, the first reasonably priced battery-electric vehicle with a range that rivals that of gasoline engines.

Alternative-propulsion vehicles still haven't lived up to their potential in North America. Their higher up-front costs remain significant for a vehicle's first owners, who tend to hold on to the car for just two or three years. But thanks to the mobility disruption, we're approaching a time when the up-front costs of alternative propulsion won't matter much. Soon, personal mobility will be dominated by transportation-as-a-service models, akin to self-driving versions of today's Uber or Lyft ride-sharing services. Fleet operators will own their vehicles throughout a 300,000-mile use cycle. That future approaches sooner than we think. And it just might spell the end of the dominance of the internal-combustion engine.

Chapter Six

CLOSE ONLY COUNTS IN HORSESHOES

The hardest thing about prizefighting is picking up your teeth with a boxing glove on.

——FRANK "KIN" HUBBARD

In October 2007, a month before the DARPA Urban Challenge, I went out for lunch with GM CEO Rick Wagoner. Rick was in a great mood. The company had just weathered a difficult period. The powerful United Auto Workers had targeted GM as the lead Detroit company for the latest round of negotiations. Topic A in the talks was what to do about GM's health care obligations. Since the sixties and seventies, when foreign automakers first began competing with American brands in the domestic market, Detroit had complained that companies like Toyota, Nissan, Honda, and Volkswagen enjoyed a competitive advantage because their factories didn't face the same union obligations that Detroit's did. For example, in 2007, Toyota's labor cost was about $50 an hour, including pension and health care obligations. GM's labor cost was $80 an hour—and that $30-an-hour difference multiplied by GM's 73,000 American workers complicated our ability to develop new and compelling products. It also added thousands of dollars to the sticker price of GM vehicles. In the previous two years we'd lost $12 billion and cut tens of thousands of jobs. Rick had had enough; heading into the negotiations in the last week of September 2007, he had directed his team to separate GM from its health care obligations. We were going to try to work with the UAW to create a health care trust, a voluntary employee benefit association, or VEBA. Essentially, GM would pay billions of dollars into a new entity, managed by the union, then forget about ever again having to pay health care costs for its hourly employees. Liberated from our obligations, we could concentrate on making money selling great cars.

The labor negotiations had been so tough that UAW president Ron Gettelfinger ordered GM workers to strike the company for the first time since 1970—and each side dug in for a long work stoppage that had the potential to cripple the company. Three days in, a late-night agreement was reached in which GM would pay $38.5 billion in cash and stock to create the health care trust. The same agreement cut in half the starting wage of new GM hourly employees, from $28 to $14 an hour. The media called it a landmark contract. "A big step forward," one pundit declared.

Inside GM, we all felt the same way. The deal had boosted our stock price by $9 a share, to a three-year high. At the next automotive strategy board meeting, I saw some of my peers as jubilant as I'd ever seen them. "This agreement helps us close the fundamental competitive gaps that exist in our business," Wagoner said in a statement. He was even more effusive during our lunch. "From here on out," Rick told me, "I think it'll be much smoother sailing." Rick's time as an officer in the company had been devoted to halting GM's long-term slide in market share and sales. Now Rick felt like not only had that slide stopped, but the company's trajectory was poised for a steep ascent. Much of the automotive strategy board shared his optimism. Now, many of us felt, we'd finally reached the point that we could rack up some sales, generate some real profits and fulfill the potential of the team that Wagoner had built. Most importantly, from my perspective, I'd finally have the budget required to bring us closer to the mobility disruption finally becoming possible.

Whatever higher power has dominion over the auto industry must also exhibit a powerful appreciation for irony. Because just as the pieces were falling into place to solve the sustainability issues that had for the better part of its history plagued the American automobile industry, Detroit faced its biggest-ever crisis. Heading into 2008, battery technology was growing robust enough to begin to make electric vehicles practical. Satellite-enabled mapping capability made it possible for cars to locate themselves on the Earth's surface. Smartphones would soon make ride-sharing services viable. Sensing technology had grown adept enough to provide a vehicle with a detailed picture of the world, and computing power was so fast that those same vehicles could drive themselves. The technology for a widespread disruption in personal mobility was ready. We just had to put the pieces together. And just as we were ready to do so, Detroit began feeling the effects of historically high gas prices and the subprime mortgage crisis. As consumer demand for automobiles fell, the engineers working inside the industry found themselves in a race against time.

That race would dominate my life for the last two years I spent at GM.

Within weeks of my lunch with Rick, Alan Greenspan referred to the "bubble in housing." Home prices fell, mortgage foreclosures rose and regular people started to grasp that Wall Street's subprime mortgage crisis might create some collateral damage in the overall American economy. But even as gas prices nudged up past three dollars a gallon, few people had any inkling of the economic tumult to come—or how it would affect the auto industry.

So when one of Rick Wagoner's classmates from Harvard Business School called him up, asking whether GM wanted to buy Segway, the personal-mobility company well known for inventing a self-balancing scooter, Rick was curious enough to give me a call about it and to get my sense of the deal. "We don't want to buy a company that makes a stand-up transporter used by someone wearing a helmet, riding around looking like a nerd," I said. "But maybe there's a way to collaborate with Segway on a different play."

I had followed Segway's trajectory for some time at that point, like most people who were interested in personal mobility. A remarkable amount of curiosity had been generated by inventor Dean

Kamen and his investors before the 2001 unveiling of his personal-mobility device, which in classic tech-boom fashion had a code name—Ginger—and some remarkably ambitious goals. Silicon Valley venture capitalist John Doerr compared the Segway's import to the Internet, while Steve Jobs said it would be bigger than the PC. Kamen himself said he believed the Segway "will be to the car what the car was to the horse and buggy." Kamen caught my attention by making some of the same points I'd been discussing for years: "Cars are great for going long distances," Kamen told *Time* magazine. "But it makes no sense at all for people in cities to use a 4,000-lb. piece of metal to haul their 150-lb. asses around town."

In the six years that had passed since Segway's launch, it was apparent that Dean Kamen's reach had far exceeded his grasp. Segway used a portion of its rumored $90 million in venture capital funding to build a 77,000-square-foot factory that was said to be capable of assembling 40,000 Segways per month. Kamen's goal was to sell 50,000 Segways in the first year; by the time we met with them in 2007 they hadn't yet shipped that amount. And yet, the company remained a source of fascination to me. I respected the way they tried to improve personal mobility. Both GM and Segway were coming at the problem of getting around from two different directions. We were known for huge SUVs that could transport at high speeds across the country a complete family, plus their luggage. Meanwhile, Segway was known for a little device that moved a single person a few blocks through a crowded urban core. I had an inkling that maybe we could collaborate to create a new product that resided somewhere in the middle.

By the time Rick and I got to talking about Segway, GM had already signed on along with its China partner, Shanghai Automotive Industry Corporation, as the exclusive automotive sponsor of the Shanghai World Expo, which would be held in 2010. GM of China's automotive group president, Kevin Wale, had asked me to create something for the pavilion that would demonstrate GM's global technological leadership. He wanted something as important as the Futurama exhibit that GM had staged for the 1939 World's Fair in

New York, which is credited with setting the stage for America's system of interstate highways.

The theme of the Shanghai World Expo was "Better City, Better Life." It was all about urbanization—which was something that had been obsessing me anyway. The planet had nearly 7 billion people at that point. More than half of them currently lived in cities, and with each passing year a greater proportion of the world's population moved into crowded urban areas, worsening traffic congestion and air quality, and hastening planet-wide climate change. I was so concerned about the problems that arose from automobile use that I'd contacted one of the world's experts on the issues that cars posed for cities, Bill Mitchell, the onetime dean of MIT's architecture school, who now ran the Smart Cities project at the college's Media Lab. By brainstorming with Mitchell, Chris Borroni-Bird and others on my staff, we struck on the idea that GM's exhibit would present some sort of a solution for the mobility problems experienced by a crowded city like Shanghai twenty years in the future. But by the fall of 2007, we hadn't yet identified exactly what that solution would be. Time was running out, and by that point, I didn't think any of our ideas were good enough to captivate the Chinese audience.

I hoped a visit to Segway might generate some new ideas. That fall, a group of GM engineers and I flew to Segway's New Hampshire headquarters in one of GM's corporate jets. Soon after we landed, we met the company's chief engineer, Doug Field, who'd worked alongside Kamen. Earlier in his career, Field had spent six years at Ford. He understood the auto industry yet had his feet firmly set in start-up-tech culture. Field also had a prototypical Midwestern vibe—unassuming, confident, straight-shooting. We got to talking about GM's Autonomy concept, and Field showed us the way the Segway Personal Transporter was built on similar design principles—one electric motor for each wheel set into a module that might as well have been our Autonomy skateboard cut in half.

At some point, we moved into a boardroom so that I could give a presentation, one that argued that technology had the potential to solve the auto industry's sustainability issues. I asked Segway's team to imagine a new type of vehicle—a pared-down human transporter that was electric like the Segway but able to transport two people around urban areas. The idea was to demonstrate the most cutting-edge technology GM was involved in developing—autonomous vehicles that could communicate with one another—and put it in some sort of a larger version of the Segway Personal Transporter. To create a vehicle intended precisely for densely populated urban areas.

Doug Field liked the idea. In fact, he said, Segway's engineers had been working on experimental prototypes that, with a little tinkering, might serve exactly our purposes.

— — —

By the time the businesspeople at Segway and GM had worked out a deal, it was near Christmas. I had a quick conversation with Field conveying how excited I was. Field was to get going on developing a prototype for the collaboration and would call me when he had something ready. I didn't expect to hear from him for many months. But I was used to the pace that predominated at General Motors. Field's metabolism was considerably higher.

At Segway they whipped up unconventional prototypes all the time. They even designated special days just to work on outlandish engineering concepts. They called them Frog Days, which earned their name because Segway founder Dean Kamen likened them to the fairy-tale princess who kisses a frog and finds a prince. Segway engineers were told to work on whatever they found fun and interesting. That's how Field approached this GM concept. He threw his top engineers at it and set them a challenge: Could the engineers employed by this scrappy little tech start-up impress the executives of the world's biggest car maker, which had thousands of engineers on its payroll?

Just six weeks later, still deep in the winter of 2008, Field called me up and told me he had something ready. I flew over to New Hampshire the first chance I got. I wasn't really sure what I was in for. I mean, how much could Field have done in so short a time?

What I saw was one strange-looking vehicle. For one thing, it had only two wheels, like a Segway. The thing had the same balancing capability as Dean Kamen's Personal Transporter. To create the concept device, Field had split a Segway scooter to provide a wheelbase that was wide enough to accommodate two people seated side by side. On a plywood base he mounted a pair of bucket car seats. The steering mechanism looked like something out of an arcade—a pair of handgrips the driver tilted forward to move ahead and back to reverse. The command to turn happened with a simple twist of the console. Around all this, Field had used white PVC plumbing pipe to construct a protective frame that supported a windscreen and the roof.

Field handed me a black bike helmet, donned a fluorescent-yellow helmet of his own and gestured for me to take the driver's seat. The self-balancing mechanism was unnerving for a minute, then came to seem completely natural. Same with the steering. It was so responsive: just a quick twist of the handgrips and the little mobility pod twisted right or left. It was tricky to keep a straight course at first, but after a few seconds I got the knack. With some practice I was able to direct it through the Segway lab's workstations. In fact, I drove it so fast that the wheels squealed on the polished concrete floor.

I was thrilled. I couldn't get over how quickly Field had worked. It would have taken GM six weeks to write up a purchase order just to buy the parts for the concept. Field had created a whole workable device that performed flawlessly through the test. I asked him to keep working, went back to Detroit and told Rick Wagoner he had to try this thing. By the time I was back in New Hampshire with Rick, Doug had refined the concept further. This time, he took us to the place where Segway conducted their high-speed testing—an old, abandoned warehouse where someone had duct-taped mattresses around the steel I-beams holding up the roof. Rick was dressed in a suit and tie. Meanwhile, a half-dozen Segway engineers in jeans and untucked dress shirts were gathered around a Rubbermaid garden shed. Doug handed Rick a personal digital assistant and showed him what button to push. Rick pushed the button, and some electronic whirring and beeps came from the Rubbermaid shed. Moments later, out rolled the little mobility pod, with no one in it. It looked much the same as the first pod that I'd tested, but this one had a metal framework holding up the windscreen, the better to protect its passengers. The bucket seats had been replaced with a pair of rocking chairs.

"Do the seats look familiar?" Field asked.

They did, actually.

Field grinned. "They're Cracker Barrel rocking chairs."

The resourceful Segway engineer had taken a pair of the little gliders that sit out front of the restaurants and hooked them into the mobility pod's controls. Now making it go was really simple: You just leaned forward in the rocking chair. Slowing it down was as easy as pushing back on the chair. The left–right steering still happened with the video-game controller mechanism. Field had boosted the battery power, giving the little pod a top speed of 35 mph. Once you got the hang of it, you could zip the vehicle around the warehouse. Whether you were slaloming through the warehouse supports or reversing into the Rubbermaid shed, the vehicle felt maneuverable and responsive in a way that was perfectly intuitive.

Rick shook Field's hand. "Good job," he said.

"The thing feels like the transportation version of an iPod," I exclaimed. What I meant at the time was that the little mobility pod was tiny and electric, sure, but also really fun and useful. Rick and I left Segway that day deep in conversation about the potential for the mobility pod. It would be perfect to form the centerpiece of the Shanghai World Expo. Later, I thought about the iPod analogy more and realized the two inventions just might share another attribute: Just as the iPod had revolutionized the music business, I wondered whether inventions like Field's stood to trigger a similar disruption in the auto industry.

--- --- ---

Meanwhile, the subprime mortgage crisis and high gas prices were continuing to make things tough for the auto industry, including

GM. The parts supplier Delphi was struggling to escape bankruptcy protection in a sales climate that looked to be the worst in ten years. GM was struggling, too. Throughout the development of our two-person transporter with Segway, I was also working to keep GM out of bankruptcy, just like everyone else on the strategy board. The board's meetings happened once a month on the thirty-fifth floor of Tower 300 in downtown Detroit's Renaissance Center. The thirteen top-ranked members of the General Motors leadership gathered around an oval conference table to debate and consider every aspect of the company's operations. Through the course of 2008, these meetings grew increasingly somber as the economic crisis depressed consumer demand and access to credit to purchase automobiles.

In 2007, General Motors had sold 9.369 million vehicles, making it the world leader for the 76th year in a row. But Toyota was a close second, having sold a mere 3,000 fewer vehicles around the world. And with gas prices climbing to $3.50 a gallon, GM's SUV-heavy product mix was looking vulnerable: Our U.S. market share slipped from 26 percent to below 23 percent in just a few months. Sales for subcompact cars like the Toyota Yaris climbed 50 percent; truck sales sank 17 percent in a month. Analysts were calling it the most dramatic shift in consumer preferences in decades.

During one strategy board meeting near the beginning of 2008, GM's chief sales analyst, Paul Ballew, provided a forecast for the year that predicted a slowdown in business. For the previous decade, the American public bought somewhere between 15 million and 17 million new light vehicles a year. In the last months of 2007, the effects of the subprime mortgage crisis had begun to slow sales. Ballew said he thought American sales for 2008 could dip below 15 million, to somewhere near 14 million, which would be the slowest sales in a decade. The following month he revised his projection still lower, to 13.5 million for 2008. That troubled all of us on the board. At that point, we were losing a billion dollars a month. Our cash reserves were down around $25 billion and sinking fast, when we needed a minimum of $10 billion on hand to finance our payments to parts suppliers. I asked Ballew whether he could see a scenario where sales dipped to 12 million per year. Ballew said that a drop to 12 million would amount to the worst downturn ever in U.S. auto sales. As it turned out, annual U.S. light vehicle sales fell to 10.4 million vehicles per year in 2009. While we didn't see it at the time, we were doomed.

Our finance staff started going over our budgets to try to find every last penny that could be diverted to meeting our operating costs. When he was promoted from CFO to president and COO in March 2008, Fritz Henderson noted that we'd cut our costs by $9 billion over the previous three years—and said he wanted to cut another $4 to $5 billion still. We slashed our white-collar staff by 30 percent, called for a temporary shutdown for many of our domestic assembly plants and halted the development of several new models. Amid all this, GM's vice president of engineering, Jim Queen, brought up my program with Segway to develop what we were calling the PUMA prototype. (PUMA was an acronym for Personal Urban Mobility and Accessibility.) "We're just going to have to shut down PUMA," he said. "We can't afford to spend money on this."

Rick did not want to cut my research budget. "We can go bankrupt by running out of money, or bankrupt by not being technologically relevant," Rick would tell the strategy board. But things were growing so dire at GM that it was politically difficult for him to be seen avoiding cutting *any* budgets. By the third quarter of 2008, our cash reserves were down around $16 billion—with bankruptcy looming when they dropped to $10 billion. I sympathized with the effort to cut costs. But I also felt the pressure to demonstrate to the American public the innovations that were then possible. To protect our development effort with Segway, I called up Kevin Wale. He told the automotive strategy board how important our work was. Not necessarily for the good of the world. No, Wale's interest was more pragmatic. We'd committed to demonstrating the vehicle we were developing with Segway at the Shanghai World Expo in 2010. If we backed out of the Expo now, GM of China would be embarrassed, and Chinese sales were certain to fall. And at that point, Wale pointed out, GM's Chinese sales were one of the company's few bright spots.

With that protection and political cover, the PUMA project was saved. Yet every day another headline yelped about some other terrible event for the auto industry. To cut costs, GM's Saturn, Saab and Pontiac brands were looking like they'd disappear forever. In fact, everything seemed dire for Detroit. Contributing to the sense that the city had entered some sort of an apocalyptic end-time was the 2008 Detroit Lions' 0-for-16 season. Mayor Kwame Kilpatrick resigned the same fall after pleading guilty to two counts of obstruction of justice, among other misdeeds. GM was forced to join Ford and Chrysler in asking the federal government for money to see us through the downturn. Rick, Ford CEO Alan Mulally and Chrysler CEO Bob Nardelli bounced back and forth from Washington as they lobbied for the bailout. The U.S. Treasury Department, congressional leaders from the House and Senate—the three CEOs approached them all. "GM, Teetering on Bankruptcy, Pleads for Bailout," ran one *New York Times* headline as shares sank to their lowest level in sixty-five years, the situation made all the more humiliating because it came just two months after the one-hundredth anniversary of General Motors' 1908 incorporation.

Talk of an auto industry bailout climaxed on November 18, 2008, when Rick joined Mulally and Nardelli to appear before the Senate Banking Committee to ask for $25 billion to keep us out of bankruptcy. The committee was not receptive, and the atmosphere in Washington grew hostile after ABC News ran a story criticizing the three CEOs for flying to D.C. in corporate jets—with GM's costing a reported $36 million. "If the CEOs want hardworking taxpayers to bail them out, they should leave the gold-plated transportation at home," read a *New York Times* editorial. The attention convinced Ford to attempt to weather the downturn without asking the government for money. The following month, in one of his last acts in office, President Bush arranged a $17.4 billion bridge loan for General Motors and Chrysler. But no one was under any illusions that the money would be enough to save either company.

The federal loan was designed to tide us over until the arrival of the next administration. A lot of people expected the next president to require new leadership of the auto companies in exchange for more cash. And if Rick Wagoner was fired as GM's CEO, I wasn't sure how much longer I would stick around.

If I left, I worried the opportunity to create a more rational, zero-emissions vehicle for cities could pass the industry by. I didn't want to miss the chance—and so I pushed for some sort of public launch of the vehicle we were developing with Segway. Luckily, my good friend Scott Fosgard was at that point running GM's auto-show presence. Fosgard observed that GM was getting killed in the press for making big, gas-guzzling cars. Perhaps this vehicle we were developing with Segway could be used to indicate that GM was, in fact, interested in being a green company? Fosgard's argument earned him a small budget to launch PUMA at the New York International Auto Show in April 2009. I knew our little two-wheeled concept wasn't completely ready to be shown to the public, but what other option did I have?

— — —

As we worked furiously to get PUMA ready for public eyes, President Obama was inaugurated. February 2009 saw the lowest sales the industry had seen in twenty-eight years. Annualized, sales for the year would have been 9.1 million vehicles. That was a 41 percent drop for the overall industry compared to the previous February, and a 53 percent drop for General Motors itself. "It's just stunning how horrible the market is now," said one analyst, reflecting the industry consensus.

GM asked for another $16.6 billion from the Obama administration; Chrysler asked for an additional $5 billion. Toward the end of March, Obama's auto task force was said to be in the final stages of determining whether to provide the money—and what conditions to set if it did. Obama announced that he'd present a formal announcement of his long-term plan for GM and Chrysler on Monday, March 30.

On the last Saturday in March, I was at home in Franklin Village, northwest of Detroit, when I received an urgent email from Rick Wagoner's administrative assistant, Vivian Costello. She said the whole of the automotive strategy board needed to be on a conference call for the following morning, at nine. On a Sunday. This sort of thing never happened. I turned to my wife, CeCe, as soon as I read the email.

"Rick's in trouble," I said.

CeCe reassured me. "Maybe it's nothing," she said. "Maybe he's going to tell all of you that you got the loan."

On Sunday morning I dialed into the conference line and could feel the anxiety among the other callers. Rick disclosed his news right after he came on the line: "I need to let everybody know that I've resigned—and Fritz is the acting CEO," he said. "And, um—I appreciated serving with all of you. I'll turn the call over now to Fritz."

I was devastated. I'd known that Rick's departure was a strong possibility. President Obama was expected to force out the leadership of GM and Chrysler as a condition of the bailout. But still I took it hard when it happened. Rick had been my direct boss for more than thirteen years. Over the last decade, I had been making twenty to twenty-five international trips a year, many of those with Rick. Not only was Rick the best boss I'd ever had, the two of us were really good friends.

By the time the call ended, I had a pit in my stomach. "I've got to leave, too," I told CeCe as soon as I hung up the phone.

"Well, don't do anything crazy," she said.

—

The New York International Auto Show was held April 10–19 in 2009. We launched Project PUMA the day before the show's press previews, on April 7, just a week after Rick's resignation.

Fosgard rented an event space near the auto show's location at Manhattan's Javits Center, and along with Chris Borroni-Bird we braced ourselves for the vehicle's debut. I had my misgivings about the way the event would play out. PUMA was anything but a finished concept. We planned for it to have connected and autonomous capabilities, so that it could be summoned from a parking garage, avoid crashes and navigate intersections and traffic without human input. But none of that had been engineered into the version we were demonstrating. With yellow caution tape decorating the perimeter of the windshield, PUMA looked every bit the experimental prototype that it was.

The day before the official launch, I headed to the *New York Times* to tell the automobile editors and reporters about the vision that PUMA represented. I'd expected some tough questions. What I hadn't expected was how hostile some of the people would be to anything that General Motors did. PUMA was a *good* thing, I wanted to point out. It was an example of GM trying to improve the world. But people were so cynical about us that even PUMA was viewed critically.

Why was General Motors doing PUMA *now*? asked one reporter, trying to get at why an automaker that had previously made the lion's share of its profits selling SUVs would suddenly pivot to the creation of two-person mobility pods so small that six of them would fit into a standard parking space. My answer, basically, was because we could. Because the technology had progressed to the point that these mobility pods were possible. And not only could we do them—we *should* do them. With a range of 35 miles on a charge of electricity that cost just 35 cents, and a top speed of 35 mph, the PUMAs could serve as the default mobility option for urban populations. They made a lot more sense than drivers sitting alone in over-engineered five-passenger vehicles stuck in bumper-to-bumper traffic.

I told them we intended to bring this to market in a few years as a vehicle that GM sold. (Later, I'd say, by 2012.) And when I was asked by a skeptical reporter how much the PUMA had cost to develop, I told him just one-half of 1 percent of our overall engineering budget, taking pains to make it clear the prototype hadn't contributed to our financial problems. "And its selling price?" One-third to one-quarter of a conventional automobile, I said.

I took the stage first thing on Tuesday morning, and I could see some skeptical looks in the audience as I launched my presentation. "PUMA stands for Personal Urban Mobility and Accessibility," I explained. "It represents the next step in the reinvention of cars for our cities . . . and addresses the major global trend of people moving to and working in urban centers."

The vision for PUMA, I said, was the mobility pod whizzing autonomously about somewhere like Manhattan while the "driver"—or, more accurately named, "rider"—did whatever he or she felt like. Sending text messages. Posting to social media. Catching up on the news. Reading. Watching entertainment or sports. "We were the SUV company, and we accept that," I said to the journalists. "We *want* to become the USV company—known for ultra-small vehicles."

By my side, Segway CEO Jim Norrod described the way the transporter was fully electric. PUMA could be operated at just 10 percent the cost of a standard car. And with a turning radius of zero, thanks to its twin opposable wheels, it was much more maneuverable than conventional vehicles.

As we spoke, I was sensitive to the journalists' reaction.

"You may be wondering why GM is working on a project like this given our current situation. When you look at the challenge facing GM over the next couple of months, and for the rest of 2009—we are putting all of our attention on reinventing the company. Project PUMA is consistent with that effort. As a matter of fact, it's right in line with what President Obama and the Auto Task Force spelled out last week—the need for GM to lead the way in developing clean technology and energy-efficient vehicles for tomorrow."

Afterward, we gave some of the reporters rides in the PUMA. The reporters seemed to enjoy the rides, but the resulting press coverage was not kind. "Wow, that is one vehicle I can safely say I have absolutely no interest in driving," said the tech website Geekologie. Another website, Engadget, called it a "rickshaw without the charm." "It looks like someone pimped a wheelchair," one onlooker said. And an Internet commentator observed that it "does the same job as a pair of legs, but without all the simplicity and ease of use." Finally, the *Washington Post* noted that PUMA was "dismissed as a gimmick, or 'not a real car,' or something that, as one young scribe put it, 'only a geek would buy.'" But that same reporter provided the rare support for PUMA. "After going nowhere quickly in a Yellow Cab, after watching cars clumped together on city streets, rendering them nearly impassable, it dawned on me that maybe the traditional automobile has reached the end of the road . . . The PUMA pod, initially expected to be sold to test-bed communities, such as college campuses, just might be the way to go."

At least we had one supporter.

By this point I'd had a few days to process Rick's departure, and I knew that I wouldn't be around to see PUMA move from an experimental prototype to something ready to be demoed at the Shanghai World Expo. I'd have to leave that up to some of the engineers I managed, like Chris Borroni-Bird. The PUMA launch would become my last high-profile public appearance with GM. Perhaps there was something fitting in that. Yes, the PUMA had received a lot of criticism at the New York International Auto Show. But I remained confident in the potential of such vehicles.

Weeks after the New York auto show, the Obama administration announced a plan that would result in the bankrupted GM closing 12 to 20 factories, cutting 21,000 union jobs and shuttering 2,400 dealerships. I couldn't get over the timing. Just at the moment that I could so clearly see that technology stood to solve many of the auto's big problems—the automobile industry was cratering. To succeed Rick, the president's Auto Task Force brought in a Texan named Ed Whitacre, Jr., the former chairman and CEO of AT&T, to be GM's new chairman. During his first appearance before the strategy board, Whitacre gave us a "win one for the Gipper"-style pep talk that irritated many board members because the tone was so out of sync with what we were feeling. By May 2009, I could feel the strain dragging me down. While I had been a daily runner for the past thirty-five years, I could barely manage a flight of stairs. One weekend I was in my house and I couldn't get out of bed. My daughter Hilary, the only person home at the time, somehow got me into the car and drove me to the hospital, where we waited in the ER. I visited the washroom, and when I'd been gone for some minutes, my daughter had someone check on me. I'd collapsed in the men's room. The doctor admitted me to the hospital's isolation ward because they couldn't figure out what I had. It turned out to be a relief to learn that my exhaustion had, amid everything else, contributed to my contracting a bad case of pneumonia.

On September 30, having carried through with my pledge to resign, I left the General Motors' tech center for the last time as an employee, feeling like I had a scarlet letter on my chest—a big *B* for bankruptcy. Who was going to be excited to hire an executive whose company had slid into insolvency? Aside from the time I spent working at my father's diner as a teenager, GM was the only employer I'd ever had. Still, I was proud of my record there. In an article the *New York Times* ran to mark my departure, the writer called me "the company's champion of electric and hydrogen fuel-cell vehicles." "It's hard to imagine the phrase 'reinvention of the automobile' being spoken at General Motors' headquarters without it being connected to Lawrence D. Burns," the story began, noting that I'd served as the chief of R&D for longer than anyone who had held that post since Charles Kettering, the man who invented the position. Later, the newspaper referred to me as a "big picture guru of sustainability."

The time off did provide me with the ability to work on the book I'd written with my colleague at GM R&D, Chris Borroni-Bird, and the MIT Media Lab professor Bill Mitchell. It was called *Reinventing the Automobile: Personal Urban Mobility for the 21st Century.*

Bill Mitchell was one of those historic reformers of cities and transportation I spoke about earlier. As the dean of MIT's School of Architecture and Planning, Mitchell had advocated for, among other things, cities that were more hospitable to human beings while still being easy spaces to move around. Later, he was in charge of MIT Media Lab's Smart Cities research group, which called for the wholesale reinvention of urban transportation. Mitchell regarded contemporary cities as broken—and the thing that had broken them was the automobile. One of his big projects was something called the CityCar, a tiny, lightweight electric vehicle, similar in many ways to PUMA, but with some important differences. The PUMA had solved the parking issue by eliminating two of a car's traditional four wheels, making it easier to fit more vehicles into the same space. The CityCar remained four-wheeled, but to make it more efficient when it wasn't in use, Mitchell designed it to fold up and fit together with other CityCars when parked, a bit like a stackable chair.

Reinventing the Automobile stated my feelings on the automobile's situation a lot more bluntly than I might have put it had I stayed at GM. The book described how much trouble cars caused the world, and argued that a new iteration of automobile represented a powerful part of the solution to these problems. Conventional vehicles were powered by internal-combustion engines, energized by petroleum, mechanically controlled and operated in a stand-alone fashion, unconnected to other cars or the world around them. The new automotive DNA, we said in the book, would employ electric and autonomous technologies in vehicles that were tailor-made for city use.

While I was inside GM, I'd often felt like no one was listening to my descriptions of the future of mobility. Once I left GM, I discovered that wasn't the case. It turned out many people agreed with me that large-scale disruption in the industry was inevitable—and they wanted my guidance on how to navigate it when it happened. Soon I was consulting for numerous corporations that were interested in the future of mobility, from trucking firms to oil and gas operations and everything in between.

Visiting Shanghai to take in GM's exhibit at the 2010 World Expo provided me with the opportunity to witness a portrayal of the future in action. Before I left GM, PUMA had evolved into EN-V, for "electrically networked vehicle," although we pronounced it "envy." To move away from PUMA's utilitarian roll-bar design and caution tape, program manager Chris Borroni-Bird had solicited ideas from GM designers around the world. With the help of designer David Rand, Borroni-Bird selected three prototypes to actually build. GM Europe created a streamlined red space helmet that the marketers dubbed Jiao, Mandarin for "pride." GM Holden of Australia had designed the Xiao ("laugh"), which resembled a robotic version of a duck's head. And the company's Advance Design Studio in California created the Maio, Mandarin for "magic," a smoked-glass-and-neon take straight out of *Tron*.

After I left GM, Borroni-Bird maintained his steady hand overseeing the PUMA project, and he'd done a wonderful job. The EN-V exhibit launched in May 2010 in the Shanghai World Expo pavilion that GM shared with Chinese automaker SAIC Motor. The pavilion depicted the little two-person mobility pods autonomously using electric motors to zip about urban areas, picking up passengers and dropping them off and moving on to the next ride. American vehicles were so big and powerful that it was hard to imagine the PUMA sharing the road with them. But in China, you could envision the little EN-V pods zipping around among the bicycles, motorcycles, mopeds and all the other transportation solutions the Chinese used.

In contrast to PUMA's launch in New York, the EN-V prototypes were big hits in Shanghai. "While these are still concepts, they show GM is thinking outside the box on making different kinds of vehicles to address concerns such as traffic gridlock and gasoline consumption," said the *Orange County Register*. "[A]n EN-V is less than half the size of a MINI," observed *The Economist*. "The advantage of having only two wheels is that the car can be shrunk into a small package . . . Automated driving, moreover, takes the EN-Vs to a new level of sophistication." Compelling design showcased in the context of Shanghai had transitioned PUMA, panned by the media just a year earlier, into the star of the World Expo.

One night in Shanghai I woke up at 2:00 A.M. and couldn't get back to sleep. Jet lag. So I put on my workout clothes and headed up to the hotel gym, where I got on a treadmill and started jogging. Below me the brown waters of the Huangpu River wound their way through the city, dividing the east and west halves of Shanghai. I couldn't get over the ship traffic. GM's world headquarters in the Renaissance Center provided a similar view over the Detroit River, where maybe a single freighter passed by each hour. Here was completely different. It was the middle of the night on Sunday, and boatloads of gravel and coal, and whatever else, passed by the hotel at a rate of about one a minute.

The world, I realized, was a much bigger place than I'd grasped from inside GM. While the U.S. was just pulling itself out of the recession, the Chinese economy was growing something like 8 to 10 percent a year. There were so many more opportunities, so many more people hustling and bustling to make a buck.

In my hotel, overlooking the river and the highways and everything else, where every possible transportation option was being exploited, Shanghai seemed like the city of the future. I felt like we were almost there. Almost at the point that we had a workable solution to Earth's mobility problems. Something was going to happen in the next several years. The current system was just too broken for it not to.

III

THE AGE OF AUTOMOBILITY

Chapter Seven

THE 101,000-MILE CHALLENGE

*Confidence is what you have before you understand
the problem.*

——WOODY ALLEN

In the immediate aftermath of the DARPA Urban Challenge, the only major step toward a self-driving car happened because a handful of San Francisco television producers couldn't get pizzas delivered to their studios.

This remains remarkable to me, more than a decade later. Given the success of the DARPA Urban Challenge, and the resulting publicity, you'd think that corporate America would have jumped on developing a self-driving car. Television shows and documentary teams covered the challenges. Dozens of articles chronicled the events. Shouldn't that have been enough to start a stampede into the self-driving-car space?

Today, I would say yes. Actually, it boggles my mind that in the immediate aftermath of the November 2007 race, no major American corporation used the momentum from the challenges to stage a serious and well-funded effort to make self-driving cars possible.

I bear as much responsibility as anyone for that fact. Think about it: I was the guy running research and development for GM, then the world's largest automaker. *We had funded the winning team!* Not only that: Carnegie Mellon's Red Whittaker and Chris Urmson asked me outright to fund a joint venture between GM and the Pittsburgh university to continue the research that had won the urban challenge. And I said no.

Why? Because at that point, in 2008, GM was devoting all its energy to surviving historically low consumer automobile demand. Nearly all of my fellow GM executives considered autonomous cars to be a half century away, at least—if they even considered the possibility at all. With the specter of bankruptcy looming, there was no way I could gather the budget required to pursue the opportunity, a fact that I regret to this day. And the other major automakers were too busy fighting bankruptcy as well.

Once I declined the GM–Carnegie Mellon joint venture, Urmson tried to start a robot car-racing league, even traveling to Qatar to enlist backers, but he ultimately got nowhere. In the immediate aftermath of the race, the only major American company to attempt to use the momentum of the Urban Challenge to do something incredible was Caterpillar, the construction- and mining-equipment giant, which put together a team from Carnegie Mellon to develop self-driving mining technology through the university's spin-off, the National Robotics Engineering Center. To get the project going, Caterpillar licensed the software that Tartan Racing had built for the Urban Challenge. Bryan Salesky ran the day-to-day operations on a multiyear project for Caterpillar to create enormous self-driving dump trucks, with Chris Urmson working under him, as well as Carnegie Mellon faculty member Tony Stentz and former Tartan Racing team member Josh Anhalt. The idea was to develop the technology until it could help run open-pit mines autonomously, with unmanned dump trucks trundling deep down into the mines, pausing by a robot excavator to receive loads of ore and then heading off, completely autonomously, to dump the loads into processing equipment.

On the other side of the country, in Silicon Valley, Stanford's Sebastian Thrun was throwing the expertise of his team at some of Google's key projects. Once they met the goal of gathering a million miles of road-level imagery for Street View in just seven months, Thrun moved them on to an even more ambitious task.

Google insiders referred to the project as Ground Truth. While the Urban Challenge was happening, bidding wars erupted for the world's two biggest digital-map makers, Chicago-based Navteq and Tele Atlas of the Netherlands. The major mobile companies were jockeying for position as they developed the capabilities of location-based search—the ability to locate products or services in close proximity to the cell phone user performing the search. (So an executive on a business trip might be able to find "best closest running trail," for example, or students on a night out might seek a nearby "24-hour place to buy a slice of pizza.") But access to Navteq's and Tele Atlas's mapping services was expensive. Google was spending millions a year to license its maps data, while having strict restrictions placed on its use. For example, one company wouldn't allow Google to provide turn-by-turn directions because the company already was selling such a service to a competitor. In the fall of 2007, while Thrun was working on Street View and overseeing Montemerlo's running of the DARPA Urban Challenge team, the navigation device manufacturer TomTom bought Tele Atlas for $4.3 billion, while Navteq went to Nokia for $8.1 billion. Google was well on its way to developing Android phones, which ultimately would debut in September 2008—and which would directly compete with Nokia. Because those phones would provide location-based services, they would also compete with TomTom. With the two maps providers in the hands of competitors, Google realized that it had to develop its own map technology—and quickly.

Thrun and his team, which again included Levandowski, would play the key role in solving that particular problem. The Ground Truth project also featured Megan Quinn, a well-known venture capitalist today, working as project lead. The name, Ground Truth, is based on a cartography concept describing the extent a map accurately reflects reality—the truth of what you see on the ground. The project began with maps provided by such entities as the U.S. Geological Service. Thrun's big innovation was the realization that Street View imagery represented a great way to aid ground truth for

Google's digital maps. Similar to how he constructed a map of the Smithsonian's interior for his Minerva robot, Thrun and his team used hundreds of cars equipped with rooftop rigs to create digital maps. Then they ran the imagery through artificial-intelligence software to capture key details, such as addresses and street names. AI was taught to recognize and understand the meaning of traffic signs to assist with turn-by-turn directions. The final part of the Ground Truth process involved human beings going over every aspect of the maps and correcting any bugs created by the AI. This could be anything—perhaps a store featured a large arrow on its sign that the AI interpreted as delineating a one-way throughway, when in fact the street was a usual two-way route. In that case, a human computer operator would select the road and correct the data. Thrun outsourced the work to several thousand technicians in Hyderabad, India. Volunteers within Google also went over the maps. Megan Quinn baked chocolate chip cookies for anyone who discovered bugs, one cookie per bug. All told, Quinn distributed eight thousand cookies.

Throughout all this, Thrun's key lieutenant was Anthony Levandowski. The former Velodyne LIDAR salesman and creator of the Ghost Rider self-driving motorcycle was the critical person in the development of the Street View camera-rig hardware, the actual mechanism that captured the road-level imagery and wedded it to accurate location information. Levandowski's juggling of interests was to raise questions later. According to Mark Harris's reporting in *Wired*, people at Google referred to the hardware rigs as "Topcon boxes," believing them to be sourced from a Japanese maker of optical equipment named Topcon. What few knew was that a company Levandowski co-owned, 510 Systems, had designed the boxes and then licensed the technology to the Japanese manufacturer. (Levandowski would disclose to Google his role with 510 Systems, as well as the fact that 510 had designed the technology.) Representing Google, Levandowski purchased technology through an intermediary from his own, external company.

At this point, Google was delighted with Levandowski's work with Thrun. Their efforts on Street View had established a cheap way to provide the illusion of 3-D immersion on locations that would stretch across the world. With Ground Truth, the pair's innovation saved Google millions in licensing fees. The accuracy and turn-by-turn direction capabilities of the map data likely generated hundreds of millions in stock market value. And Levandowski was a key part of all this.

In addition, he was developing inertial measurement units for the Street View vehicles, managing 510 Systems and flying back and forth to Hyderabad to work with the army of techs correcting the Ground Truth maps. According to Harris, he was also creating stock-market-prediction software with another of Thrun's acolytes, Jesse Levinson. Between 2007 and 2009, it seems, Levandowski was a whirling dervish of productivity and innovation. As it happened, it also was left to Levandowski to take the next step in pushing forward self-driving technology.

—————

In early 2008, a Discovery Channel television show called *Prototype This!*, which chronicled the exploits of inventors attempting to create ambitious innovations under a two-week deadline, approached Levandowski with an unusual proposition. The show's production company had an office on Treasure Island, a man-made landmass in San Francisco Bay linked via an isthmus to Yerba Buena Island, which in turn was linked to the mainland by the San Francisco–Oakland Bay Bridge. The unusual location seemed glamorous, but for the people who worked there, it did have one drawback: Few restaurants would ever deliver takeout to their offices because they were technically outside of San Francisco's borders. The problem gave the producers an idea for a *Prototype This!* episode: Challenge inventors to come up with new ways to get food—specifically, an order from the legendary San Francisco establishment North Beach Pizza—to the island, and film the results. One inventor proposed firing the pizza over to the island by a rocket-powered rail gun; another, delivery by blimp. In February 2008, Levandowski suggested transporting the pizza from San Francisco over to the island with a robot car.

Thus, the first significant use of the technology created for the DARPA Urban Challenge was a TV publicity stunt motivated by man's hunger for pizza. To develop the driverless vehicle, Levandowski first used a setup very similar to the one he'd used for Google's Ground Truth project, featuring a rooftop mapping rig that employed a LIDAR scanner, multiple cameras and a GPS unit, as well as a wheel-hub device that took ultra-precise measurements of the rotation of the vehicle's wheels. With the rig activated, Levandowski drove the route from North Beach Pizza's Stanyan Street location near Golden Gate Park onto I-80 and across the Bay Bridge to the show's Treasure Island studio. The mapping rig noted the location of every tree, every house, every office tower. Driving the route created a 3-D scan of the territory that Levandowski's robot car would have to navigate.

Levandowski chose a silver Toyota Prius to equip with the LIDAR, radar and computer equipment required for autonomous function because, he said, the drive-by-wire system was relatively easy to hack. In addition to the self-driving-car equipment, the Prius, which Levandowski dubbed Pribot, was emblazoned with the logos of its sponsors, such as 510 Systems, Topcon and a new company, Anthony's Robots. The television producers also arranged to equip it with a key piece of custom technology: a trunk-mounted storage locker specially designed with insulation to keep the pizza hot.

The morning of Sunday, September 7, 2008, was a clear day, free of precipitation, the perfect conditions to test an autonomous vehicle. The show's producers arranged a sizable police escort— both cruisers and motorcycles. They also had to shut down to all other traffic the roads the robot Prius would travel. That introduced a layer of complexity to the task. The producers were only able to arrange a shutdown of the Bay Bridge's *lower* level, which placed a significant amount of concrete and steel between the Prius's GPS module and the satellites the module would be trying to contact, making the GPS unavailable during that portion of the trip.

The Prius, reportedly the first autonomous vehicle ever to be tested in San Francisco, followed its motorcade of police vehicles and filming equipment along the Embarcadero without issue. It slowed along the Bay Bridge and proceeded safely and calmly past the last receipt of its GPS signal, using the LIDAR-created map to navigate the lower bridge deck. Through it all, *Prototype This!* host Mike North and Levandowski rode ahead of Pribot on the back of a flatbed truck. "Anthony, you realize you're making history right now?" North said to Levandowski at one point.

The trickiest spot involved the exit ramp off the Bay Bridge onto Treasure Island—a hairpin left turn, in which the Prius turned too sharply and wedged itself against the leftmost wall. It was later discovered that the accident happened because Levandowski's team had forgotten to program into the software the correct dimensions of the Prius. Levandowski was forced to clamber off the flatbed and slither in through Pribot's open driver's-side window to take control of the car and free it. Once liberated, the car continued autonomously onto Treasure Island, transporting the pizza to the *Prototype This!* warehouse. "That's a topic like—crazy," says Thrun, who still marvels a decade later at Levandowski's achievement, calling it "a completely historical event" for the way his colleague, in just weeks, and on a comparative shoestring budget, was able to create a vehicle that drove in a populated area without a human being inside.

— — —

Because the robot stuck itself against the Bay Bridge's exit-ramp wall and needed an assist, it didn't quite complete the first fully autonomous pizza delivery—but let's give Levandowski credit for the first autonomous crossing of the Bay Bridge. In retrospect, his achievement was significant less for its firsts and more for what came after. Pribot and the pizza delivery happened as Sebastian Thrun completed work on Ground Truth. Thrun and Larry Page were having an ongoing discussion about what to do next. "Look, Sebastian," Thrun recalls Page telling him, "you should work on self-driving cars."

"Why?" Thrun asked.

"Well," Page said, "if this business succeeds, it could be bigger than Google. Which means, even if there's just a ten percent chance of this succeeding, it's worth the investment."

But in the direct aftermath of the DARPA Urban Challenge, Thrun was not convinced that self-driving cars were a near-term possibility. The Victorville event had happened in an empty city, bereft of pedestrians, and even then, Thrun thought, the vehicles had to creep around the race course and occasionally ran into things. Thrun said, "The performance in the Urban Challenge did not make me think, wow, we were on the brink of a new revolution."

So he told Page, essentially, "It's too hard. It can't be done." But Larry Page came back the next day. "I've thought about this. Give me a technical reason why it can't be done. Not a societal reason. A *technical* reason."

Page was looking for something definitive. Something like, today's computers aren't yet powerful enough to process the data required to safely navigate a highway at legal speeds. That wasn't actually true—but that's the sort of reasoning Page sought. Rather than supplying him with the technical reason, Thrun became a bit frustrated with his boss. This time, he didn't just say it couldn't be done. He said, "*Damn* it, it can't be done!'" Recalling the conversation a decade later, Thrun says he reminded Page that Thrun was the world authority on autonomous vehicles. "The world's best professor in this field," Thrun recalls. "Not those exact words. But, I was like, 'Trust me—I'm the expert.'"

And still Page wouldn't take no for an answer. He came back to Thrun the following week. "Hey," the Google founder said. "Just tell me—I want to tell Sergey and Eric that it can't be done. But they would love to have a technical reason."

So Thrun went home and tried to think of a technical reason. And he couldn't. "Shit," he said to Page the next day. "I can't really think of a technical reason."

"Maybe it *can* be done," Page said.

At this same time, Thrun's colleague Levandowski was busy

creating on a meager budget an autonomous vehicle that ended up navigating from downtown San Francisco across the Bay Bridge. The thought is certain to have occurred to Thrun—what sort of technical feats might be possible with the sort of budget Google's backing could provide?

Soon after, Thrun sent out an email invitation to what he considered the best minds that had competed at the DARPA challenges, so that the group of them could get together and consider whether and how to make a serious effort to commercialize self-driving-car technology—to make autonomous automobiles a reality. Today, a decade later, Thrun credits Page with convincing him to pursue the self-driving-car project—all of it set against the backdrop of the development of a pizza-delivering robot car.

In October 2008, a month after Levandowski's pizza delivery and nearly a year after the DARPA Urban Challenge, Thrun gathered together an A-team of robotics brainpower at his Lake Tahoe chalet.

From the Stanford team that competed in the challenges was Thrun's old friend, the mapping expert, Mike Montemerlo. Dirk Haehnel and Hendrik Dahlkamp were German software engineers who had worked with Thrun on Junior, Stanford's entry to the Urban Challenge. Levandowski was there. So were the technical director and the software lead of the Carnegie Mellon DARPA challenge team, Chris Urmson and Bryan Salesky. All the men but Thrun, who was then forty-one, were in their late twenties or early thirties.

The gathering was designed to answer two questions. The first was whether to work together at all. A positive answer to the first question provoked a second issue: How to do it? Google was willing to fund the effort. Did it make sense to conduct the effort from within Google or as a separate start-up?

There was a lot of reminiscing that weekend about the DARPA challenges. The weekend also likely featured wide-ranging discussions about the proper approach to pursue developing autonomous technology in the real world. Over the years lots of different variations on autonomous vehicles had been suggested by various futurists. Some involved sensors and other infrastructure that were incorporated into roadways, to communicate with vehicles. So a stop sign might be implanted with a radio-frequency chip that announced its presence to approaching vehicles. Little tags implanted under dashed yellow and solid white lines might announce the center and sides of the road. But the self-driving experts barely considered such possibilities. Building out such infrastructure would require a massive effort. And what might happen if one or many of the sensors stopped working? A bug in a chip that signaled a stop sign could have disastrous consequences. As a group, the engineers and programmers believed the best course to real-world autonomous cars involved putting all the intelligence in the vehicle.

At one point during the weekend, Urmson and Salesky headed off in their rental car and drove the winding roads around Tahoe to discuss Thrun's proposal by themselves. By that point it was sounding as though the effort would be a Google thing. Salesky wasn't sure about it. He thought of Google as a search-engine company. Why would some Silicon Valley Internet outfit bother to throw money at developing driverless cars? It seemed so tangential to the company's business.

Salesky argued further that he and Urmson already were working on a tremendously exciting project. Their Caterpillar-funded initiative to create autonomous dump trucks would get self-driving tech into the hands of a manufacturer with a track record of scaling complicated machines at high volumes. As roboticists, Urmson and Salesky lived to create robots that would affect regular people. And the Caterpillar project, which was already in operation, stood to do that, Salesky argued. "I just thought the mining problem was easier to solve," he recalls.

Urmson had a different take. The Caterpillar project was fascinating work, but it involved creating robots for *mines*. The best they could hope for, in the current project, was for Caterpillar to create a few thousand dump trucks that would operate in remote operations around the world. The work would incrementally make it cheaper to mine ore. It might incrementally reduce the number of lives lost in mining accidents. But so far as regular people were concerned, it wouldn't revolutionize anything.

Google's project, Urmson argued, could transform the world. Could transform the way billions of regular people around the planet went from place to place.

Urmson also saw Thrun's pitch as an opportunity for his own career. Salesky was leading the Caterpillar work, and Urmson had leadership ambitions of his own. Thrun wasn't just getting this self-driving-car team going. As he tended to do, he maintained other responsibilities. He hadn't yet fully extricated himself from running Ground Truth. And he also was talking with Larry Page and Sergey Brin about running a new entity for them, an offshoot of Google intended to pursue other audacious, high-cost and high-reward projects that had little to do with Google's core search business. Thrun's other responsibilities would make it difficult for him to manage the day-to-day operations of the self-driving project, Urmson realized. When the group was still talking about pursuing the project as a start-up, Thrun had proposed that Urmson would be the new entity's CEO. Urmson was bound to occupy a similar leadership position if the project happened under the Google umbrella. Leading the best financed effort to work on a project that had an opportunity to change the world was a chance Urmson couldn't pass up.

The timing was awkward for Urmson. With his PhD work complete, he was in the process of accepting a position at Carnegie Mellon's Robotics Institute as full-time faculty. At the meeting to introduce new professors to their fellow academics, Urmson had to step forward and explain to everyone that, actually, he was going on sabbatical for two years to work with Google. It was a tough blow to the Robotics Institute—the loss of one of its top young minds, a star from the DARPA challenge teams. But it was probably hardest for Salesky, who opted not to join the Google project, preferring instead to stay behind in Pittsburgh to continue his leadership of the Caterpillar project. Salesky wasn't just losing a valued colleague— he was also losing one of his best friends.

— — —

After the meeting in Lake Tahoe, Thrun hired a team of about a dozen engineers to staff the project, which became known within Google by the code name Chauffeur. Each one of them were experts in a particular skill that was integral to the autonomous operation of an automobile. Almost all of them had participated in the DARPA challenges. The team included Chris Urmson, the project's engineering lead, who was charged with developing the driverless car's software. Levandowski, who had handled the sourcing and development, and the actual building, of the hardware for Street View, would hold a similar role for Chauffeur. Thrun's right-hand man on the DARPA Urban Challenge, Mike Montemerlo, had the responsibility of developing the maps the autonomous vehicle would require to locate itself in the world. Former Stanford DARPA Urban Challenge team member Dmitri Dolgov, a Russian programming expert, would work closely with Urmson to write the planning and control systems that operated specific aspects of automobile operation, such as braking and steering. Street View veteran Dirk Haehnel would work on software infrastructure. Principal software engineer Jiajun Zhu would help craft the perception system. Another Street View veteran, Russ Smith, would build an algorithm that used wheel rotation and the car's own inertia to track its movement in space. And Nathaniel Fairfield would create the simulation software that would enhance the car's safe operation by running it through hundreds, thousands and ultimately millions of situations in the virtual world.

From the beginning, the self-driving team was focused on achieving two audacious but clearly defined milestones. The approach hatched by Brin, Page and Thrun was modeled on what had worked with Street View. (And it bears repeating here that Thrun based his Street View approach on techniques he'd adopted from Carnegie Mellon's Red Whittaker.) "Let's make sure there is a milestone, and a payment of a good chunk of money if we get to that point," Thrun says, recalling the discussions he had with Page and Brin. The men went back and forth over what that milestone would be. "Look,

Sebastian," Page said. "I want this thing on any street in California to drive one hundred percent autonomous."

But that seemed too amorphous a goal to Thrun. How would the team prove they could drive every road in California, aside from going out and actually driving every road in the state? So Page and Brin devoted a day to tracing out a series of challenging drives. The idea was to create a small series of trips that would feature every form of difficulty California roads could throw at a human driver: All the bridges in the Bay Area. Downtown San Francisco. The coastal highway from Mountain View to Los Angeles. The mountainside switchbacks around Lake Tahoe. And Lombard Street in San Francisco, reputed to be the world's most crooked street. "That kind of stuff," Thrun says. "It includes everything. Cities, highways, everything. They said, if your team can drive that, then you get a good chunk of money. As a bonus."

"Let's just say that it was more money than I ever thought I would make, ever," Urmson says.

All told, the challenges that Page and Brin created for Thrun's team amounted to ten drives that totaled about a thousand miles. The handle the team gave to this aspect of the challenge was Larry1K. Thrun didn't know whether Larry1K was possible. So he negotiated a second milestone, just in case, which would also provide the team with a payout: the accumulation of 100,000 miles of autonomous driving on public roads. "Because I wanted to be sure, if we weren't able to do that one thousand miles, there would at least be some way to get some money."

Each person on the team grasped that this was a once-in-a-lifetime opportunity. It also seemed a little surreal. They'd joined a supergroup of robotics greats working feverishly against time on a secret project to achieve a goal set by two of the world's richest men—and if they achieved it, they'd become wealthy themselves. It seemed like something out of a movie. Most of them had toiled on this same problem as penniless grad students, scrimping together enough money to survive, racking up credit card balances and other debt, and deferring the pursuit of more lucrative engineering jobs.

Some of them had children, and wives who had compromised their own dreams to support their efforts. Now success on this Google team would provide them with the opportunity to make it all up to their loved ones.

Plus, it was a really cool engineering challenge. "For an Internet company that is super-well-capitalized to say, let's give this a go?" said one engineer. "That was a 'holy shit' moment for all of us. Everyone was like, whoa, let's get out there and do something magical."

It would take a few years before people understood that the project made perfect sense for Google to pursue. Realizing why requires going back a few steps in Google's story, to Street View, which at the time seemed to many like a zany science experiment with few real-world applications. Being able to zap yourself, using your computer or handheld device, to any address in the United States (and eventually, many other countries in the world)—that seemed cool. But what was the point? Why bother going to the unimaginable effort required to collect end-to-end immersive imagery of *every single road* in the United States? Google's development of the Street View service stands as a cartographic achievement unmatched since the days of Vasco da Gama and Magellan. The scale of the task boggles the mind. Not just for the way it required hundreds of drivers transporting camera-topped cars to navigate every road in the United States, Canada and most of the developed world, but also for what it led to.

The Street View project made possible Ground Truth, and excellent and accurate location-based search, because Google could use artificial intelligence and computer-vision software to pull from the Street View data the real-world location of every retailer, restaurant and address in any area it serviced. The data was useful for more than that, however. It also could be used to provide people with navigation directions. And still another use for those ultra-accurate three-dimensional scans of every road in the United States?

Autonomous cars.

That's right: Street View and its successor program, Ground Truth, represented an important step in the development of autonomous-driving technology—because Chauffeur's approach to the safe and reliable deployment of self-driving cars required highly accurate maps of the roads its vehicles navigated. By the time Chauffeur started, maps and the capabilities they provided to robot cars had already become Mike Montemerlo's specialty. "Having map data let you do some *amazing* stuff," Montemerlo explains.

First, high-resolution and ultra-accurate 3-D map data provided the Chauffeur vehicles with the ability to locate themselves in the world, just as preexisting maps did for Thrun's Minerva robot tour guide and Levandowski's pizza-delivering Pribot. The general method was similar. Vehicles equipped with mapping equipment, such as LIDAR and cameras, would drive the same area numerous times throughout the course of several nights and days, scanning everything as they did so to create their own 3-D model of the world. By comparing the things that changed position between scans with those that didn't, the software could create a list of stationary landmarks, like curb edges, telephone poles, homes and buildings, mailboxes and billboards. Later, when the self-driving car navigated the same territory, it would conduct a similar scan of the area around it and attempt to match its current surroundings with the list of landmarks in its 3-D maps. By aligning its current model of the world with the previous scans, the vehicle could locate where it was with a high degree of certainty—down to a margin of error of a couple of inches.

The car would also compare its preexisting list of stationary objects with the objects around it to discern which ones were likely to move—helping the car differentiate between a mailbox and a pedestrian, or a light pole and a cyclist. If a slim, cylindrical shape was in exactly the same spot on multiple scans of the environment, then that shape was likely to be a light pole. If that shape showed up only on the most recent scan, then it was possible the shape was a human being and needed to be watched.

The 3-D maps also could help the vehicle discern the dotted or solid lines in the middle of the road, with what the Google engineers called a "lane-level" map, which featured corridors that were safe for the car to proceed along, as well as lanes that other vehicles or moving objects were likely to navigate. Maps also help in cases where road lines have become obscured, thanks to dirt or snow. "The maps let you detect obstacles more reliably," Montemerlo says. "They let you figure out what's off the road versus what's on the road."

Finally, preexisting maps also provided the Chauffeur vehicle with an ultra-reliable way to detect tricky but important parts of the world around it—like traffic lights.

"For instance, if you know there was a traffic light there yesterday," Montemerlo explains, describing a capability that a great map provides to an autonomous car, "there's very likely a traffic light there today. And so instead of searching the whole image for traffic lights, you can concentrate on specific places and make your detector much more accurate. That's *super*-important."

Identifying traffic lights is difficult for a self-driving car forced to rely exclusively on its own sensors—which is one of the reasons that traffic lights weren't required in the DARPA Urban Challenge. "Buses have red lights on the back of them," says Chauffeur software engineer Nathaniel Fairfield. "Does that mean it's a traffic light? *People* have real problems with this. If you're driving through an area you don't know? Sometimes people get confused. You see a green light, but you want to turn left. Should you be waiting for an arrow? If you don't know the situation it can be really hard to interpret."

Detecting traffic lights, and correctly discerning their meaning, becomes a lot simpler for an autonomous car if it can consult a preexisting, high-resolution 3-D scan of the road that it's traveling—one that tells the car exactly where it should be looking.

"And then you can figure out the *semantics* of that traffic light," Fairfield explains. "What does it mean? Which lanes does that traffic light control? Does it control the left-hand turn lane? It's an *enormous* assistance."

Mapping every traffic light, every stop sign, every left-hand turn lane, sounds like an enormous undertaking—and it is. But bear in mind that Street View and Ground Truth had already performed similar scans of the same roads. With that effort proven possible, Montemerlo, Fairfield, Urmson and the rest of the team decided to assume that it could also be possible to map every traffic light on every one of those roads.

Now to actually build the robot. This was Levandowski's job, and he did it similarly to the way he'd built Pribot for the Treasure Island pizza-delivery challenge. He used four radar-based sensors, which bounced sound waves out from each corner of the car to sense short-range obstacles. The car had a GPS module to establish where it was in the world, and wheel encoders that tracked the vehicle's motion in ultra-accurate fashion. The required computer equipment, which had once filled the back of Boss's Chevy Tahoe, was now small and portable enough to fit into the space formerly occupied by the Prius's spare tire. But the heart of the system was the Velodyne LIDAR sensor, which resembled a spinning Kentucky Fried Chicken bucket on the roof of the vehicle. Velodyne's LIDAR was a 64-beam laser that conducted about 1.5 million measurements per second, 360 degrees around the vehicle.

For the software that would dictate every element of the vehicle's behavior, the team began with the code Thrun's Stanford team wrote for Junior. That required lots of tweaking. In one early innovation, the team's coders rigged up a system that controlled the Prius via smartphone. Tilting the phone forward accelerated the car; tilting it back braked it. Building up and testing the autonomous capability of the Prius required sitting in the vehicle for hours at a time, writing code and then uploading it to the Prius, then watching how it drove based on the tweaks. The team passed hours in this manner—code, drive, code, drive—in the upper, gravel parking lot of the Shoreline Amphitheatre, just down the parkway from the main Google campus in Mountain View. Dolgov in particular would work with the vehicle in the parking lot late into the night. One early morning, around 3:00 A.M., Dolgov was sitting in the Prius tweaking the code when he looked up and noticed an approaching police car. "How are you, Officer?" Dolgov asked the policeman in Russian-accented English, suddenly aware that the self-driving project, a project so black and confidential that very few senior people in Google even knew it existed, was in danger of being exposed.

"What are you doing?" the police officer asked, and Dolgov just smiled and explained that he was testing technology for Google. Which, thankfully, in Mountain View was enough to mollify the officer's curiosity.

Teaching a robot car to drive required learning all over again the way humans operate automobiles: for example, the precise way that humans depress a brake pedal. Most drivers don't just slam on the brakes as hard as they can. Rather, they steadily increase brake pressure, easing the vehicle into a deceleration that grows more intense until the moment before the vehicle stops, when the driver suddenly eases off. Dolgov replicated a similar curve in the Chauffeur software.

Then there was the problem with the thousands of tiny adjustments that every driver makes as he or she navigates a city street. Drivers don't simply stay in the center of a lane; they meander within it based on all sorts of external cues. Parked cars on the right side of the road force a shift to the left. Oncoming vehicles in the opposite lane prompt a shift to the right. Such minute adjustments happen numerous times a minute, and are required to make the vehicle's occupants feel safe.

One of the encouraging early runs of the self-driving Prius happened at nearby Moffett Federal Airfield, just a few miles east of the Google campus. Once the Prius was cruising well on an airplane runway, Urmson decided the vehicle was ready to head out on a public road. But was that even legal? The company's lawyers investigated the state's driving regulations and discovered that, at that point, nothing explicitly prevented a computer from operating an automobile. The key part of the legislation required a human to be in the driver's seat—but didn't specify that the driver had to actually be controlling the vehicle.

In May 2009, Urmson chose Mountain View's Central Expressway for the first test on a public road. Central Expressway was perfect because it was straight, well maintained and featured a wide boulevard separating the self-driving car from any oncoming vehicles. To safeguard the public, Urmson arranged for a number of cars to follow behind the Prius, to form a barrier between it and regular traffic. With Dolgov in the passenger seat, Urmson pulled the car onto Central Expressway, activated the self-driving software and found the car wiggling and shuddering all over the lane. Every little landmark on either side of the pavement—light pole, mailbox, parking sign—triggered a feint to the right or the left. It turned out, basically, to be wandering within the lane *too much*.

It was the sort of thing that no one had bothered to notice during the DARPA challenge efforts, because those autonomous vehicles weren't made for human occupants. But a comfortable and safe ride for humans was essential to Chauffeur's success. Dolgov made some adjustments to the code—essentially, preventing the Prius from moving about in its lane except in response to the closest of stationary obstacles. The next day they set off again. The tweak worked; the driving felt a lot more comfortable. "We crossed the one-mile mark, and then the ten-mile mark. That was a pretty big deal," Urmson recalls.

Then came the hundred-mile mark, and the thousand. That spring of 2009, it became apparent that they were avoiding doing the hard stuff—the Larry1K set, the hundred-mile drives that Page and Brin had designed to test the vehicle's capabilities. "At some point we said, we need to go try it," Urmson recalls. The first one they thought they'd try was the drive down the Pacific Coast Highway, from Carmel-by-the-Sea to San Luis Obispo. On the surface, this looks like the most dangerous of the lot, with sections that wind along road blasted from cliffs, where one side is a vertical rock face and the other features a mere steel ribbon of guardrail—the only thing preventing a car from plummeting hundreds of feet into the Pacific Ocean below.

But actually, Urmson figured the challenge was relatively easy for the team to pick off. The road was well maintained, the lane markings clearly labeled. Long stretches of the drive were straight with relatively low traffic. The route didn't have many intersections, which meant just a handful of traffic lights. So long as the car stayed between the yellow and white road lines, Urmson figured, they'd be okay. "The first [drive] was basically just stay in your lane, stop for a couple of traffic lights and keep going," Urmson says, while acknowledging that a mistake could have represented a potentially fatal tumble down the cliff into the Pacific Ocean. "That was both fun—and terrifying."

Once the route had been mapped by a vehicle equipped with a scanning rig, Dolgov ran the data through software designed to establish the precise path the Prius should take along the way. That is, at each point in the road, the computer program specified the spot where the self-driving vehicle should be. Next, Dolgov and his wife, Anna, a member of Chauffeur's test and operations team, headed out to follow the drive down the coast. Along the way, as Dolgov drove, Anna corrected the path the software calculated the vehicle should take. If the pre-identified optimum position seemed too close to this or that edge of the road, Anna had only to press a key on a laptop to correct it to within ten centimeters to the right or the left.

Urmson, Dolgov and Levandowski were in the Prius the first time the robot tried to drive the route. They set out in the morning. Just a few miles in, it quickly became apparent that they needed to do something about the code that regulated braking on curves. Dolgov had written the software so that the car didn't brake until the point where the curve became tightest. The Prius felt like it was heading into the curve far too quickly, then slamming on its brakes and creeping through the latter half—an unnerving tendency on a road that wound high above the Pacific shore. Urmson's knuckles whitened with each switchback and S-turn. But the three men resisted disengaging the autonomous software: It would be great for morale to pick off the first of Chauffeur's ten challenging drives on their first attempt. Eventually the Prius came up behind a slow-moving beer truck, and the truck's braking in the curves slowed down the autonomous vehicle as it was forced to maintain a cushion behind. Well down the coast, the Prius came upon a construction project that reduced the road to a single lane, regulated by a worker directing traffic. Humans signaling when to stop and go represented one of the trickiest challenges for self-driving vehicles. Faced with the added complexity of the route, Urmson disengaged the autonomous capability and assumed control.

They didn't give up. Rather, they returned to the start. Dolgov tweaked the code that regulated braking into curves so that the software told the Prius to decelerate some time before the car reached the road's point of greatest curvature. Later that afternoon they tried again. The vehicle's tweaked curve-handling felt much safer. And by the time they reached the location of the previous disengagement, the construction project was done for the day. The road was back to two lanes, without a signalperson managing traffic. On the second try, they made it all the way to San Luis Obispo. It was a winding, cliffside drive of about 132 miles, which took more than three hours. And the Chauffeur team's vehicle had done it completely autonomously.

They'd picked off the first of the Larry1K routes. "We talked about how big a deal it was," Urmson recalls. "How you could start to see how this might work . . . It was more excitement than relief."

That evening, he, Dolgov and Levandowski stayed overnight in San Luis Obispo. They went out to a restaurant and celebrated. "We'd set these milestones," says Dolgov, describing chest bumps and high fives. "But we didn't know how to get there—and we'd just knocked off the first one. It was just another thing that fueled the momentum and the excitement."

The pessimists on the team dismissed the achievement. Big deal, they said. The road was simple. Picking off that one required only that the vehicle stay in its lane and navigate just three traffic lights. Actual city driving was far more complicated, particularly when you have to account for pedestrians and pets.

So Urmson sought out a tougher challenge for the second go-round. He chose the Larry1K drive that basically followed one hundred miles of a historic road called El Camino Real, Spanish for "the royal road," which in the pre-American period connected the twenty-one Spanish missions that stretched from Baja California to the northern tip of San Francisco. The Larry1K version of the drive started near the San Jose airport. In Mountain View, the route left El Camino Real to wind through the Google campus. In Palo Alto, it cruised through Stanford University's campus and downtown, and then, north of there, returned to El Camino Real, which the route then followed south of San Francisco. *This* route featured more than two hundred traffic lights. The thing that most worried Urmson about this one was driving through downtown Palo Alto. The angle-parked cars that backed into the road, the pedestrians distracted by their phones, the cyclists who zoomed through red lights: each represented a variable that human drivers found tough to predict. The idiosyncrasies of this route would be even more difficult for an autonomous vehicle. "I remember people being like, this is a waste of time, why are we doing this?" Urmson says. "But it turned out that going and trying it was the best thing we ever did."

The Prius made mistakes a half-dozen times as it drove through downtown Palo Alto. The car failed to yield to a pedestrian and to another car backing up into the street. Each time Urmson had to disengage the autonomous capability and take over, and in the aftermath Dolgov and the other programmers would work to fix the problem, correcting the code so that next time the car *would* yield to the pedestrian and slow to a stop when a car backed out into the street. "It moved from an abstract set of problems to, 'Okay, we did this, and here are the four points where it failed. Let's put our effort into solving those problems,'" Urmson says. "These are some things we can go fix. We don't have to fix *everything*. We just have to fix *these* things, and then let's go run it again."

The team met every Monday at 11:30 A.M. to discuss their progress and where to concentrate their efforts the following week to improve the car's self-driving capability. One problem they had to tackle was the software's inability to predict the future path of pedestrians. For instance, even if a pedestrian stood at an intersection to the right of the vehicle, the vehicle might proceed because it wasn't able to predict that the pedestrian could be about to walk out in front of it. Also the car tried to stick close to the road's centerline, which made things difficult when the lane was so wide it could accommodate two vehicles alongside each other. Coming up on intersections, human-driven vehicles kept pulling up alongside the Chauffeur vehicle before it could get over to make a turn. So another tweak allowed the car more leeway to float within the lane, depending on the context. Proceeding in this fashion, successfully completing Larry1K's El Camino Real route took about a month.

Another of the challenges they did relatively early was driving over all the bridges in the Bay Area. This one was tricky not because of the bridges, which ended up being fairly navigable, thanks to their clearly defined lane markings, but because it included a devilishly narrow section in Belvedere Cove on the north side of San Francisco Bay. There, a street called Beach Road, which begins near the San Francisco Yacht Club, heads up along a hill and ends up tracing the shoreline of the bay through a canopy of trees and rhododendron bushes and other vegetation, separated from the water only by hidden multimillion-dollar homes. Cars parked on Beach Road despite its narrowness. The first time Dolgov and Urmson drove on it they were confused—surely it was just one way? With the cars parked along it, the road could accommodate only a single vehicle. But numerous takes revealed that Beach Road allowed traffic to proceed in two directions. To navigate it safely when faced with an approaching vehicle, the programmers had to teach the Prius a technique, common in Europe, of pulling to the side at wider parts of the road and then waiting for the approaching car to pass. Later, during an attempt when the robot successfully passed the Beach Road section, they confronted another problem: bridge tollbooths. The route they'd mapped for the car specified the Prius should go through a certain tollbooth—which, on that day (bad luck), turned out to be closed. Luckily, that particular tollbooth was open on the next attempt, and they passed that part of the Larry1K challenge, too.

One of the high points from that year was the race the team staged for approximately fifty senior Google executives. It happened in the deserted upper parking lot of the Shoreline Amphitheatre, a short bike ride from the Google campus. The Chauffeur team created a winding race course with traffic cones and then timed the autonomous vehicle as it went through. Then they disengaged the self-driving software and handed the vehicle over to the Google executives and challenged them to drive it themselves, along the same route, faster than the robot had. No one could. The incident underscored to the team the extent to which they were creating a vehicle that could drive not only as well as a human being—but better than one. Faster, sure. But more importantly, more safely, with less of a tendency to get distracted or confused.

The real world featured things that had been barred from the DARPA course. Things that sometimes didn't behave all that rationally. Pedestrians, for example, who were apt to step into the street without checking for cars because they were text messaging a babysitter. Cyclists who were apt to execute a sudden left-hand turn into traffic. Pets that liked to dart into the road.

So the Chauffeur team had to take the set of things that moved in the world and teach the self-driving car to recognize them based on their camera and LIDAR scans. A human pedestrian tended to be anywhere between about two and seven feet tall, maybe a foot across at the torso and with legs that tended to change positions. Fed hundreds of thousands of different images of pedestrians, artificial intelligence software could use learning algorithms to identify them with a high degree of reliability, in much the same way that your phone's photo software learns to recognize the faces of you and your friends. A similar routine taught the software dozens and even hundreds of different classes of objects that represented potential obstacles for the Chauffeur vehicle. A person in a wheelchair. Toddlers. Children. Skateboarders. Dogs. Cats. Soccer balls. Basketballs. Bicycle-riding ice-cream vendors. The technology learned to recognize each one.

One of the trickiest of the computer vision problems was recognizing a traffic cop directing traffic and understanding the idiosyncratic way the cop might wave to indicate it was okay to proceed or hold up a hand to tell the vehicle to stop where it was.

All of these examples involve perception—the robot's ability to see the world, and understand what's in it. The next step beyond perception makes use of something called a behavioral engine, which predicts what all those moving things in the world are going to do in the future. For example, an adult is likely to stay on sidewalks and obey traffic signals, while a child is far more likely to jaywalk and dart into traffic. The behavior of other cars was also important. Say the autonomous Prius was in the left lane on a divided highway that featured two lanes in its direction—and that it was approaching a merger of those two lanes into one. And not only that—that a pickup truck was ahead of the Prius in the right lane. The self-driving software had to recognize that the truck was likely to merge into the Prius's lane, forcing the robot to slow down to avoid a collision.

According to Urmson, the vehicle's behavioral engine grew so accurate that it was able to predict the motion of most things around it ten to twenty times a second. Say a cyclist headed along the right side of the road but up ahead was a parked car that blocked the cyclist's path. The robot had to know that the cyclist was likely to come into the left-hand lane, the robot's lane, to get around the parked car—and that the robot once again had better slow down to avoid hitting the cyclist. Another tricky one? A cyclist in another lane of an intersection pushing through a yellow light is likely to progress through the crossing, even though the light has turned red, which means the robot had better wait to proceed *even though its light is green*. All told, the team had to teach the self-driving car to recognize thousands of different behavioral rules.

— — —

By spring 2010, the Chauffeur team had developed a routine: Most often, Urmson and Dolgov were the pair who would actually be in the vehicle when it tried to pick off one of the challenging drives. (Although others did handle the occasional route. For example, Chauffeur roboticist James Kuffner, now the CEO of Toyota Research Institute-Advanced Development, was in the Prius when it successfully drove from the Google campus down to Los Angeles' Hollywood Boulevard.)

To allow the other members of the Chauffeur team to watch, the passenger team set up a system that tracked the vehicle along its route—a bit like the visuals on software for such ride-hailing services as Uber and Lyft, with one important difference: The Chauffeur software indicated whether the car was operating autonomously. The nearer the Prius was to the end of each route, the more Chauffeur engineers followed the vehicle's progress. When they did pick off a challenge, the Prius riders would return to their offices and crack open a bottle of champagne to share with the rest of the team. Afterward, they signed the champagne bottle with the name of the challenge. The bottles accumulated on a shelf in a prominent spot in the Chauffeur headquarters.

One day, Urmson, Dolgov and Dolgov's wife, Anna, headed to Lake Tahoe to work on the two Larry1K routes in that area. The Prius picked off the first of the routes without trouble. Then they had a decision to make. They would soon have to return to the Bay Area because Dolgov was supposed to pick up his parents from San Francisco International Airport. But the three of them had some time before they needed to leave. In Tahoe, they opted to set the Prius on the second route, just to see how far it would get. Winding along mountain roads, it performed so well that Dolgov found himself in the strange position of almost hoping for a disengagement. But no—the car ended up navigating the entire second route, meaning they'd picked off two in one day, and meaning also that the three of them had to hightail it to the airport immediately after, egregiously late to pick up Dolgov's folks.

By the summer of 2010, eight bottles were on the team's trophy shelf. Then nine. As summer 2010 gave way to fall, just one route was left. And that's when Sebastian Thrun received a call from the *New York Times'* veteran technology reporter, John Markoff.

— — —

Put yourself for a moment in Chris Urmson's place. You sweated blood working on DARPA's challenges. On Red Whittaker's teams, you sacrificed time with your young family. You put off finishing your PhD, put off pursuing a career outside of academia, because you wanted to develop a robot vehicle, because you realized the enormous potential of autonomous cars to transform the world. And then, finally, when you won the best-publicized DARPA challenge of them all, nothing happened. For a time. The 2008 recession meant the auto companies and anyone else who might be interested in developing autonomous vehicles were too busy fighting off bankruptcy. So you shelved your dreams and joined a project constructing a robot designed to help dig ore out of the earth. Until Sebastian Thrun calls you out of nowhere and provides you an opportunity that seems too good to be true: working on a project for a couple of dot-com billionaires that allows you to pursue your dream of transforming the world with robot cars. And which, if successful, will make you wealthy, too, in the process fulfilling all the dreams you ever had back when you were pulling all-nighters in the Pittsburgh winter for Red Whittaker.

And in 2010, as summer gives way to fall, your Chauffeur team has already achieved the lesser of the goals, entailing the autonomous driving of 100,000 miles on public roads. As well as nine out of ten of the Larry1K drives. You are so close to achieving the milestone. And then comes the potential disaster that puts the fulfillment of your dream in jeopardy.

To accumulate the 100,000 miles, Chauffeur had built numerous copies of robot cars and hired and trained dozens of drivers to sit in the autonomous vehicles while they navigated the roads around Mountain View. For whatever reason, one of these drivers called up John Markoff with a tip about the project. With that staff member's help, Markoff contacted a second driver, who confirmed the existence of the Chauffeur project. Markoff wrote a feature, and before publishing, he called up Thrun for comment. "I hear you guys are doing experiments with robots on public streets," Thrun recalls Markoff saying during the call. Markoff sent Thrun a rough outline of the story. The gist of it was, according to Thrun, that Google was performing these tests on Silicon Valley roads. "It was a terrible story," recalls Thrun, referring not to the quality of the reporting (Thrun admired Markoff's abilities as a tech writer), but to the story's expected effects on public attitudes toward the Chauffeur project. Would the public think it reckless? Irresponsible? Thrun forwarded Markoff's email to Google PR and Brin and Page.

The Chauffeur team wondered what would happen once the news got out. "We were worried—we didn't know how it was going to break," Urmson recalls. The story as Markoff had written it seemed likely to play out badly for Chauffeur. A PR backlash could affect the Google share price. The Chauffeur team worried the entire project would be cancelled. "In retrospect, knowing Larry and Sergey better, that was a dumb concern," Urmson says. "They're above that."

While Thrun stalled Markoff and debated the Google response with the company's PR chief, Rachel Whetstone, and the leadership team, Urmson and the Chauffeur engineers leapt into action. If Chauffeur completed its milestone before the *New York Times* ran the story, then at least the engineers were certain to achieve this phase of the goal.

On September 27, the last Monday of the month, Urmson and Dolgov climbed into one of Chauffeur's Toyota Prius autonomous vehicles to make an attempt on the final challenge. The Prius had a third passenger that day: Andrew Chatham, a Chauffeur software engineer who was along because he hoped to witness history firsthand. Back at the offices, the rest of the team tracked their progress on computer monitors.

The first part of the route headed to Palo Alto and ascended

Page Mill Road into the arid scrub of the Santa Cruz Mountains, quickly becoming a winding, two-lane affair, well marked with clearly defined edges, nevertheless tricky because of its switchbacks and blind curves, as well as its sport bikers and cyclists, and the odd tech millionaire attempting a new speed run in the latest sports car. At Skyline Boulevard, Page Mill became Alpine Road, and there the fog closed in. And while the road here was less traveled, the Prius riders were alarmed to see strange objects in their path that the self-driving car had to swerve to avoid. A rusted bicycle. A men's shoe. For such random artifacts to appear out of the fog, the riders started feeling that fate itself was conspiring against them to prevent the completion of the last Larry1K challenge—until, finally, material-izing from the wispy cloud cover was a junk truck piled high with detritus, so overburdened it was dropping its cargo.

Some of the hardest going was in the town of La Honda, where the narrow roads, steep hills and vegetation would have challenged a human driver—but the Prius handled it all, no problem. Out of town they followed Pescadero Creek Road, which took them west down the other side of the Santa Cruz Mountains. A right turn brought them onto Stage Road, which narrowed into a strip of un-marked pavement that wound through treeless hills until it merged with U.S. 1—another tricky moment, considering the Prius had to accelerate from a stop onto a highway with a 50 mph speed limit.

In a neat bit of synchronicity, this was the same road they'd also navigated on the first challenge route, more than a year before. The section they drove last time was the Pacific Coast Highway from Carmel-by-the-Sea to San Luis Obispo. This time they joined U.S. 1 well north of Carmel, on a stretch known as the Cabrillo Highway. They took it north, from San Gregorio along compara-tively straight pavement through the town of Lobitos, past the leg-endary Mavericks surf off the shore of Half Moon Bay, up through a section of Route 1 near Pacifica that was known for rock falls and closures due to landslides, a section the state of California was then spending millions to circumvent with the construction of the Tom Lantos tunnels.

No landslide or rock fall would impede the Chauffeur team's progress that afternoon in September 2010. Through it all, Urmson and Dolgov watched their vehicle perform without problem. Rather than the white knuckles that had marked Urmson's first ride, thanks to Dolgov's unnerving braking algorithm, this time they cruised along smoothly and safely.

In Fairmont, the vehicle navigated a cloverleaf to exit from the Cabrillo Highway and take Skyline Boulevard past the San Francisco Zoo to the Great Highway, which lined the beaches of San Francisco's western shore. Next onto Point Lobos Avenue, past the city's Cliff House restaurant and Geary Boulevard, and then a left onto 34th Avenue and through Land's End along Legion of Honor Drive to El Camino del Mar. The Prius then made its way up through the Presidio near the Golden Gate Bridge to wind a complex route through the tourists and crosswalks, finally exiting the park at Lombard Gate. It next headed west along Chestnut, on the final few miles now, and *of course* the Google founders put it on the Larry1K. Given the geography and circumstances, how could they not? The Prius rolled up the hill of Lombard Street until the little robot car reached the crest at Hyde and then headed down the so-called "crookedest street in the world," with its tourist pedestrians and its zigzag squiggles; still, the robot made it through. "It sounds hard," Urmson acknowledged later. "But because it's a one-way route down, Lombard wasn't as difficult as it seems."

Some of the trickiest elements were on the final part of the route that Larry Page and Sergey Brin had mapped out eighteen months before. The last leg turned south up Leavenworth, jogged to the right, left, right, left onto Gough Street to the thoroughfare that slashed diagonally through the city: Market Street. It featured every complex issue that an urban area could present to an autonomous vehicle: cyclists, buses, complex lane markings, San Francisco's signature streetcars. Each, individually, the autonomous software could handle. But here, they were compressed together in one hard-to-predict environment. Would a cyclist collide with the vehicle? Would a pedestrian step out in front of the car without warning? Given Urmson's history with robots, the Sandstorm roll-over, the H1ghlander rollover, Boss's strange GPS problem with the Jumbotron—Chauffeur's chief engineer could be forgiven for being braced for some terrible, last-minute accident.

Nothing like that ended up happening. Back in Mountain View, the rest of the Chauffeur team barely breathed. Dolgov, Chatham and Urmson kept silent inside the vehicle, as though any speech might mess up its functioning. But the Prius navigated the worst that Market Street could throw at it. Through the Castro, on toward Twin Peaks, where Market veers crazily past Y-merges and some of America's most complex intersections, until Urmson, Dolgov and Chatham, and the robot, reached the end of the street, which also happened to be the end of the Larry1K challenge.

The Chauffeur team had done it. They'd met the Google found-ers' challenge—with three months to go before the two-year dead-line. "It was *unbelievably* fast," Thrun says.

Thrun threw a party for the team at his home in the Los Altos Hills—and the engineers on the Chauffeur team threw him in the pool to mark their accomplishment. Urmson and his wife put a down payment on a house with their share of the resulting bonus. Meanwhile, Rachel Whetstone, Larry Page and Sergey Brin decided to work with the *Times'* John Markoff, in the hope that educating Markoff about Google's self-driving-car effort would prompt the reporter to take a different approach to his story. In fact, they of-fered Markoff an even better scoop than he had—the opportunity to become the first member of the press to ride in the Chauffeur vehicle.

That's when I return to this story. On the ninth of October 2010, I was in Midland, Michigan, where I had just finished an after-noon of consulting for Dow Corning, when my mobile buzzed and I found myself talking with Urmson. It had been at least two years since Chris had last been on my radar. Since then, GM had gone bankrupt. I'd left the company and developed a growing career as a consultant and educator working with schools such as University of Michigan and Columbia University.

During the first part of our conversation, Urmson brought me up to speed on Chauffeur, and the way the team of engineers had accomplished the challenges Page and Brin had set them. All that had been conducted in secret, Urmson said. But now, the *New York Times* was preparing a story. The company was looking for a respected automotive industry insider to speak to the press about the potential for autonomous technology, to lend gravitas to their effort. Urmson was hoping that I would be that person.

Meanwhile, I was still digesting Google's accomplishment.

"Where on earth did you get one hundred thousand miles of driving experience?" I asked, thinking that it must have happened at a proving ground, or maybe a place like the Nevada Automotive Test Center.

"We did it on public roads," Urmson said.

That threw me. I had so many questions. Even if it was technically legal, GM never would have conducted self-driving-car tests on public roads. The lawyers would never have allowed it. They would have been too concerned about liability in the event of an accident.

Then Urmson brought the conversation back to the most pressing of the topics: Would I consider helping out Google's self-driving-car team by acting as a kind of auto-insider consigliere?

"Chris, I'd be happy to do that," I said. "But given what you've told me, I'd also love to come out and have a ride in your car."

Moving quickly, Urmson and Thrun arranged for me to visit Mountain View. Later the same day that Urmson called me, Thrun posted an entry to the Google blog announcing the project to the world. "Larry and Sergey founded Google because they wanted to help solve really big problems using technology," he wrote. "And one of the big problems we're working on today is car safety and efficiency. Our goal is to help prevent traffic accidents, free up people's time and reduce carbon emissions by fundamentally changing car use.

"Safety has been our first priority in this project. Our cars are never unmanned," Thrun continued. "While this project is very much in the experimental stage, it provides a glimpse of what transportation might look like in the future thanks to advanced computer science. And that future is very exciting."

The same evening, the *Times* posted Markoff's story on the web; it ran the next day in the newspaper. "With someone behind the wheel to take control if something goes awry and a technician in the passenger seat to monitor the navigation system, seven test cars have driven 1,000 miles without human intervention and more than 140,000 miles with only occasional human control," Markoff wrote in a positive story headlined, "Look, Ma: No Hands." "One even drove itself down Lombard Street in San Francisco, one of the steepest and curviest streets in the nation."

The story shocked America. Zipcar cofounder Robin Chase called it "a shot heard round the world." *Wired* magazine's Tom Vanderbilt called it "a terrestrial Sputnik"—an analogy I like for a lot of reasons. The Russian earth-orbiting satellite redefined what people thought science and technology could achieve. It also triggered the space race.

Google's self-driving car would also prompt its own version of the space race in the automobile industry. But not at first. At first, automakers scoffed. "A self-driving car?" a lot of people in Detroit said. "What's the point of that? People *like* to drive." But those people hadn't ridden inside a driverless car. They didn't know how relaxing the experience was. Or how remarkably the self-driving car redefined personal mobility.

My own visit to the Google campus happened the month following the *Times* story, at the end of November. I gave the Chauffeur team a presentation about the potential for a new DNA for personal transportation predicated on autonomous technology, as well as the increased capacity of lithium-ion batteries and fuel-cell vehicles—and a tailored design that employed two-person mobility pods for intra-city travel based loosely on *Reinventing the Automobile*, which had just been published. But the high point of the visit, for me, was my first ride in the Google car. I was struck by the advances in equipment that had happened in just the three years since the Urban

Challenge. Boss's computers and sensing equipment had required every bit of the storage space provided by the Chevy Tahoe's roomy interior. In contrast, the Chauffeur team's Toyota Prius just looked a lot more like a normal vehicle—aside from a few sensors and the spinning LIDAR on a frame atop the roof, and the large red button mounted on the console—the red button was what I'd push, one of the engineers explained, to resume control of the vehicle if something alarming happened.

I was nervous as we drove off the Google campus. When I was head of GM R&D and planning, I'd driven a lot of prototypes and knew well how unreliable they could be. My last experience, riding in Boss, had not exactly been pleasant. My palms were sweating as I drove the Prius onto U.S. 101, one of the nation's most crowded highways. Minutes later we were cruising a little faster than 55 mph, just another vehicle in the traffic flow.

"Are you ready?" one of the engineers asked.

Similar to the way any driver engages cruise control, I pressed a button that activated the Prius's autonomous capabilities. One of the engineers warned me to keep my hands hovering over the steering wheel. There was a pit in my stomach. My mouth was dry. Never before had I been so aware of the slender cushion of space that existed between vehicles on a freeway as they hurtled forward at speeds achievable by man on the horizontal plane only in the last hundred years. And here I was, entrusting my life to a machine.

The Prius sped forward with the flow of traffic, keeping perfectly in its lane. A gentle curve approached and I tensed, but the steering wheel twisted a few inches in the correct direction to keep us centered between those dotted lines. Then the rounded curves of a VW Beetle zoomed past on the left and swerved into the cushion the Prius had maintained between us and the car ahead. The jerk came out of nowhere. I hadn't even seen the car in my side mirror. Had I been able to anticipate the Beetle cutting us off, I probably would have hit the red button on the dash and taken over control of the car. I hadn't, though, and by the time I processed what was happening the Prius slowed, gently, until a cushion existed once again between us and the speeder in the Beetle ahead.

"Wow," I said. "That was really impressive."

The Prius had anticipated the Beetle's behavior. It had predicted the cut-off before it had even happened. A good human driver has his eyes on the road ahead and conducts checks around the vehicle as often as possible. But the Google car had sensors all around it. It knew what was happening ahead, as well as to the right and the left, and behind—at all times. The Google car had eagle eyes on the road everywhere, and the intelligence to let the VW cut in front of us.

Once I processed what had happened, I relaxed. A little later on, we came up alongside a tractor-trailer, and the Prius responded. It shifted in its lane a few inches to the left, away from the tractor-trailer, and I realized this vehicle wasn't just programmed to obey the rules of the road. No, it was far more sophisticated than that. It was acting like a human driver, but much better. There's a dance that humans do on freeways, or whenever they're behind the wheel, and the Google engineers had taught this car to dance.

In the Chauffeur Prius, I felt at ease behind the wheel on a California freeway like I'd never felt before. Thrun, Urmson, Levandowski, Dolgov and the rest of the team had taken something that had worked only under certain test conditions at the DARPA Urban Challenge and made it viable for the real world, where the driving hazards were more complicated by an order of magnitude: Animals. Cyclists. Pedestrians. Other drivers liable to cut you off on the interstate. Pulling back into the Google campus I not only realized that these guys had achieved their goal, but saw, for the first time, the actual extent to which their achievement was going to change the world.

Toward the end of my visit, Sebastian Thrun and Chris Urmson approached me about joining Google full-time. Commercializing the technology—that is, getting the autonomous capability into millions of vehicles the world over—would require working closely with an established automobile manufacturer, and perhaps several. The team could use an auto-industry insider, with contacts in Detroit, who really understood the implications of autonomous vehicles.

I absolutely wanted to be a part of what Google was doing. Larry Page and Sergey Brin were the only two senior executives *in the world* to put up the money required to attempt to bring this technology to regular people. Chauffeur had achieved something remarkable in my eyes: The project had accelerated the development of autonomous vehicle technology by a good ten years.

So of course, I was thrilled by the opportunity to contribute. But by that point, I'd already committed to numerous consulting projects. Plus, relocating wasn't an option for my wife's career. So instead I pitched Sebastian and Chris on my becoming a consultant part-time to the company. That suited them, and I began working with the Google team on January 1, 2011.

Chapter Eight

THE SEEDS OF CHANGE

Everybody has a plan until they get punched in the mouth.

—MIKE TYSON

The members of the Chauffeur team would recall that period of 2009 through 2010, when their all-encompassing focus was devoted to meeting the challenges set out for them by Page and Brin, as their honeymoon phase, when everyone cooperated on a single goal and everyone was getting along. The project was secret in those days, and sometimes Urmson and his team imagined how society would respond to the news of their invention. Using lasers and artificial intelligence, chutzpah and engineering know-how, they had built a car that could drive itself on public roads more safely than a human driver. They believed their invention would have a transformative effect on society.

Then the honeymoon ended. And it did so for a lot of different reasons. One was just human nature. We all do this—we look forward to an event with a collection of hopes, dreams and fears, and when the event actually happens, the reality doesn't live up to our expectations. No one threw a ticker-tape parade for the engineers on the Chauffeur team to celebrate their achievements. The world did not greet the news that Google was developing a driverless car with rapturous applause. The money was nice. That made a big difference in their day-to-day lives. But money aside, very little changed after the team reached its milestones and John Markoff's *Times* story was published.

The reception that Chris Urmson and Anthony Levandowski received in Detroit is illustrative of society's response as a whole. Once they recovered from the party at Thrun's house, the Chauffeur engineers began discussing with Thrun and the Google leadership how exactly they would commercialize the self-driving car. Most of the possibilities discussed involved working with an automobile manufacturer. So before they'd settled on any one project, Urmson and Levandowski flew to Detroit to talk to several of the major automotive companies, as a kind of investigative foray. "The idea was, if you're going to make self-driving cars, you have to work with a car company," Urmson recalls. "Maybe they'll sell us cars to build a fleet. Maybe we're going to be retrofitting our stuff onto their cars to sell. But you need to have some kind of a relationship with them to do it properly."

The first meeting was with what's known as a tier-one supplier—a company that provides parts to the brand-name automakers. These are located all over the world. The best known are Germany's Robert Bosch GmbH, Japan's Denso and Canada's Magna International. Headquartered near Detroit are companies like Lear, Delphi and Visteon. The biggest of them have revenues in the billions, with just as many employees as the automakers they're supplying—and headquarters that are just as impressive. At a boardroom in a Detroit suburb early in 2011, Urmson and Levandowski gave a presentation about the Chauffeur project, the cars' capabilities, the number of miles they'd driven, the broad strokes of how the autonomous software saw the road. And at the end, they were greeted with complete and utter disinterest. Years later, Urmson paraphrases the response of the senior tier-one executives with an overt shrug of his broad shoulders. Why is Google bothering to do this? one executive asked. It doesn't make any sense, said another. Maybe in fifty years, said a third. They seemed to think it was ridiculous that Urmson and Levandowski would ever contemplate anyone selling a vehicle with a spinning LIDAR on the roof—because to the executives, that looked ridiculous. Not one of the auto executives in the room believed an autonomous vehicle could make a commercial product anytime soon. "They just kind of laughed and thought it was cute that we were doing this," Urmson recalls.

During the same trip, Urmson and Levandowski went to visit one of the major U.S. automakers. In a boardroom with that company's senior leadership, Urmson ran through his presentation, and the response this time included some anger. "They basically thought it was incredibly reckless that we would go do this testing," Urmson says. "We'd been very thoughtful and careful about how we tested. And these guys, without really understanding what we'd done, or how, said, 'How could you possibly do this kind of testing? What were you thinking?'" The executives implied that it was morally reprehensible for Chauffeur to have brought these robot vehicles out on public roads, because doing so endangered human lives. "And it's never going to work," Urmson recalls the executives saying. And then, they saw the engineers out the door.

Urmson and Levandowski went to Detroit to say, basically, wouldn't it be great to do something together with this amazing technology we've developed? And the response they got was pretty much a flat no. Even worse, there was a patronizing tone to Detroit's reaction that would complicate relations for years between Chauffeur and Detroit. A suggestion that the way they'd solved the problem was just not the way things were done in the auto industry. "It didn't make sense to them," Urmson recalls. "And it seemed so far out of the playbook that it wasn't even addressable."

And so Urmson and Levandowski headed back to Detroit's Metro Airport. They got in their rental car and there was silence for a little bit. "Well," Urmson finally said. "I guess we're not working with those guys."

— — —

In the beginning months of 2011, as I began working with the Chauffeur team, I confronted many examples of the way Detroit's culture, the culture in which I'd existed for more than thirty years, was different from the way things operated in Silicon Valley.

I was excited about my new consulting assignment, but I was also aware that I was a fifty-nine-year-old former executive from a bankrupt car company. Before I reported for my first day in January 2011, I was concerned I'd stick out, or be regarded as a curiosity. In the weeks leading up to my first day, my fish-out-of-water anxiety focused on what I should wear. GM's dress code was business casual. Open-collared shirts, khakis, sweaters and blazers were all OK. Whenever you hosted an important visitor, you were expected to wear a coat and tie. I did this a lot, so my wardrobe was mostly business suits, dress pants, white shirts and ties. The dress was considerably more casual at Google. No suits. No ties. Sometimes, not even pants. Sometimes they padded around their workspaces in shorts and bare feet or Vibrams.

It took me about a year to feel comfortable working in Mountain View. Ultimately, I realized the secret was to just wear whatever I wanted—which tended to be khakis, a dress shirt and a sweater. Because no one else cared what I wore. In addition to being very intelligent, creative and motivated, the Google employees were accepting of people with different backgrounds.

The differences between Silicon Valley and Detroit were much larger than just the dress code. People brought their dogs to work. I loved that—I have two dogs at home, and I always missed them when I traveled. Others brought in their infants. In fact, I once almost stepped on a baby napping in a carrier next to a workstation.

Another difference, which pertains more to the problems that Urmson and his team had attracting any interest in cooperation from Detroit, was that Google was full of people who were open to new and interesting ways to solve the problem of transportation. Urmson rode his bike to work. Sebastian Thrun famously would wear Rollerblades all day to get from place to place in the office. The area around Silicon Valley was full of public transport options, like the Bay Area Rapid Transit trains and San Francisco's streetcars. And Larry Page and Sergey Brin had distributed colorful bicycles all over the Google campus to provide any employee with the means to quickly get from one place to another.

For the senior executives of the automotive companies, most of whom lived around Detroit, there was exactly one way to solve the transportation problem: with a car—one that you drove yourself. Detroit was run by car guys who derived daily pleasure from the operation of their automobiles—the thrill that came from depressing the accelerator of a powerful engine or steering through a curve in the road. It was extraordinarily threatening to people who have grown up in the auto industry to contemplate taking the driver out of the equation.

"This is never going to happen—people *like* driving," industry executives would tell me, again and again.

"I *know* there are people who like driving," I would respond. "You know what? There were people who liked to ride horses."

It was somewhere around this period that Thrun happened to attend a conference that also featured Alan Mulally, then the respected CEO of Ford Motor Company—who now occupies a seat on the board of directors of Alphabet, the Google holding company. Thrun introduced himself and mentioned that he oversaw Google's self-driving-car project. "There was absolutely zero interest in talking to me," Thrun says. "I know Alan quite well, now, from him joining Google's board. He's a nice gentleman. A very smart individual. But it just was clearly not on anyone's radar screen. If I talked to Detroit CEOs and told them I was working on a self-driving car, they would give me a smile and walk away."

That February, Fiat Chrysler began airing a television commercial for the 2011 Dodge Charger that summed up Detroit's general response to self-driving technology. The commercial begins with a shot of distant headlights driving toward the camera through a tunnel. "Hands-free driving," intones the voice of *Dexter* star Michael C. Hall. "Cars that park themselves. An unmanned car driven by a search-engine company."

Then there's a pause.

"We've seen that movie," Hall says. "It ends with robots harvesting our bodies for energy." And as the muscle car accelerates past the camera, the voice concludes, "This is the all-new 2011 Dodge Charger. Leader of the human resistance."

———

Fiat Chrysler also happened to be on the short list of auto companies that Google sought to work with to commercialize its self-driving technology. That Fiat Chrysler was depicting the Chauffeur project as something like the vanguard of Skynet, the intelligent-machine network from *Terminator*, did not bode well for any large-scale development deals in the immediate future. Within months of Google going public with its self-driving-car effort, Detroit had positioned itself as opposing any future that involved autonomous vehicles. That was a big letdown for Urmson and the Chauffeur team, who now faced a development cycle much more difficult than they had initially envisioned. The culminating reflection of this oppositional stance happened when Urmson invited a senior General Motors representative to come out to Mountain View to take a ride in one of the vehicles, and the executive spent the entire ride sharing his negative take on the experience. He literally could not have been more condescending. The attitude underlying his general demeanor was a sort of supercilious amusement, suggesting that the Google engineers were no more than freshmen and sophomores working on a high school science project—which had no prospect of ever being developed into a commercial product bought by consumers. "I'm sorry," said the guy—and I'm paraphrasing here. "But I just don't get the point. Why even bother developing this?"

The criticism of the project was frustrating in the extreme. We felt like we were working on something that would change the world, that would have a positive effect on many of society's most pressing problems, from pollution to the most basic challenge of just getting around our planet. And here was Detroit saying, basically, *Hey, this is neat stuff—but we have to be careful. We have to protect people from this irresponsible Silicon Valley company endangering people with these robot cars.* I was hot about that. The industry's products bore culpability for the deaths of 1.3 million people a year in auto crashes, and a company comes up with an invention that could end all that—and all Detroit could do was criticize the effort. Detroit's response struck me as incredibly irresponsible. I thought this self-driving-car technology was so transformative, in terms of safety, efficiency, the automobile's environmental effects, that Detroit's appropriate response should have been a rapid embrace. Those auto companies should have been falling all over themselves to work with Chauffeur.

———

As if the team needed another indication that the honeymoon phase of the project was over, for the first time they became riven with a personality conflict that would harm their ability to work together. Setting the stage for the conflict was the fact that Sebastian Thrun had less time to devote to the running of the team. Thrun had been around about a day a week during Chauffeur's first phase. But soon after Chauffeur began, Brin and Page promoted him to something they called "Director of Other"—an executive position whose responsibilities involved research and development into ideas that had little to do with the company's core search business. The following year, the job would evolve into director of the top secret Google X lab. As head of Google X (which, under the reorganized Alphabet entity, is now referred to only as "X"), Thrun was in charge of the company's attempts at what was known as "moonshots." The label is a reference to Kennedy's sixties-era attempt to redefine the space race by putting a man on the moon before the end of the decade. It describes an idea that is so outlandish it sounds crazy—and which might just work.

Thrun's portfolio at Google X, located in a series of three-story buildings at the edge of the company campus, soon included not just the self-driving-car project, but also an attempt to solve the problem of widespread Wi-Fi access with high-altitude balloons. The group delved into research developing a cable that stretched from earth to outer space, known as a space elevator because it was intended to launch satellites. Another project investigated the potential of contact lenses that monitored the body's glucose levels, which would be helpful for diabetics.

With Thrun's attention increasingly on the running of Google X, Urmson and the team worked to determine what Chauffeur would do next. Thrun refers to this period in the first half of 2011 as "the phase where we decide, Okay, let's go for real." But "for real" at what? The Chauffeur engineers sought something that could become a commercial product. Should they sell it as an aftermarket option, a retroactive roof rig, that people could purchase at a retailer to equip their own cars with autonomous capability? Deploy it somehow in large-scale private developments, like retirement communities in Florida? As a test case, the team even worked for a time on a small self-driving taxi service that ran for about a month on Google's campus, using golf carts. To make that service useful it needed to be able to cross a major road. But detecting approaching cars required a long-range sensor, like LIDAR, which was too expensive to install on a golf cart. The Chauffeur team considered erecting a sensing rig that would sit permanently at the Google campus's crosswalk, but again, that was discounted as too expensive considering the slight benefit the service would provide to the company's staff.

Such uncertainty worsened the team's internal divisions. "I think this happens to sports teams, too," Urmson says. "Everyone's united, you get to the end of the season, win the championship or whatever—you had a mission, and now it's over. So what's next?" Throughout the previous two years the engineers had worked together well as each team member subsumed his individual ambition to pitch in and do what was necessary to meet the twin challenges. That was rational. Doing what benefited the team increased the likelihood that the whole group would receive the payouts promised by Google when they met Page and Brin's challenges.

But as Chauffeur debated various projects through the first months of 2011, the engineers began thinking about their own interests—which, for a time, led to a political work environment rife with interpersonal conflict. Senior Google membership became convinced that the existence of 510 Systems, which had been involved in developing inertial measurement units for Street View and Chauffeur, and drive-by-wire technology for Chauffeur, represented a conflict of interest for Levandowski—albeit a disclosed one. It also was possible to conceive of a future where 510 Systems became a potential competitor to Chauffeur. So Google began discussing acquiring the company.

One iteration Thrun considered involved elevating Levandowski to a role akin to CEO of Chauffeur. But numerous members of the team protested that they would leave the team if Levandowski led it. The uncertainty led to Thrun sending an email to the whole of the team that March, to focus everyone on the greater goal. "We have climbed our first 4000m mountain, and now we [are] looking at [an] 8000m mountain," Thrun wrote. "We have not yet saved a single life. We have not yet enabled a single blind or disabled person to operate a car. We have not saved a single gallon of gas . . . I want to make sure we don't forget the essentials of Chauffeur: We rock, and we will change the world. And we will do this as one happy team."

Another possible deal that Levandowski championed involved key members of the self-driving team leaving Google, all together, so that they could pursue the development of autonomous cars on their own. One form of that deal involved the team joining Levandowski's 510 Systems. Yet another option discussed was the creation of a new start-up. Levandowski was so persuasive that he convinced three of the key software engineers to decide to leave. The three programmers went to Urmson and told Chauffeur's chief engineer that they'd negotiated with Levandowki on Urmson's behalf. They'd convinced Anthony to make Urmson the CEO of the new entity. And not only that—Urmson would be rewarded with a large chunk of the independent company's equity. But Urmson didn't agree that leaving Google represented the best way to commercialize the technology. "My model was, Google pays very well," Urmson recalls. "We built this thing here. This is going to take a lot of resources to go build, so [Google] seems like the right place to do it."

That was the start of the difficulties between Urmson and Levandowski. The three engineers decided to stick with Urmson and continue working for Chauffeur, within the Google umbrella.

Levandowski seemed to hold that against Urmson. "Anthony was not happy that we didn't leave with him on his master plan," Urmson says. "That was the heart of it."

Also in play was the creation of another set of milestones designed to motivate the Chauffeur engineers. There was some discussion of pegging a reward to the incorporation of the technology into 100,000 vehicles. Making that difficult to finalize was the fact that Google's leadership and the Chauffeur team hadn't yet decided their next objective. So rather than any single milestone, Google came up with the idea of creating what came to be known as the Chauffeur Bonus Plan, which would guarantee the engineers a significant share of Chauffeur's equity for twelve years after the plan's finalization, with payouts timed in intervals around every four years. The idea was to provide the engineers with the sort of payout they might get if they created a successful start-up that went public, or was bought out by another company—but to do it under the umbrella of Google. "We were focused on getting a start-up-like compensation system," Larry Page would say later in court documents. "And start-ups pay people a lot of money if they do something significant."

In the Chauffeur Bonus Plan, Levandowski secured a bigger share of the payout than Urmson, in part because Google had decided to purchase 510 Systems to ensure that the Chauffeur team's autonomous-vehicle intellectual property stayed under Google control. Some believe that part of the purchase price was rolled into Levandowski's share of the Chauffeur Bonus Plan. It would be years before Urmson and Levandowski discovered exactly the amount they would receive from the plan, because the deal was tied to Chauffeur's valuation as a stand-alone company. I never spoke of the bonus payouts with Levandowski or Urmson—but knowing what I know of the two men, it's hard not to imagine Levandowski's bigger share not rankling Chris Urmson.

Yet another factor in the tension between Urmson and Levandowski was Levandowski's close relationship to Sebastian Thrun. By 2011, Levandowski had been working closely with Thrun for twice as long as Urmson had. Even before the three of them ever collaborated on Chauffeur, Levandowski had supported Thrun through some intense times, including the period in March 2007 when Thrun was negotiating with venture capitalists to invest in the mapping company that Google ultimately would acquire and roll into the Street View project. "I think Sebastian sees some of himself in Anthony," speculates Urmson. "They're very similar in many ways. They're both very smart, they're both very entrepreneurial, they're both very high energy. They're very similar people . . . They clicked better."

Levandowski also had a habit of sending long emails to Larry Page, sharing his thoughts on Chauffeur strategy, which often contradicted Urmson's own views. Page, in turn, valued Levandowski's contribution to Chauffeur, sending an email in 2011 to make sure the Chauffeur Bonus Plan would "make Anthony rich if Project Chauffeur succeeds."

Meanwhile, Levandowski used his relationships with Thrun and Page to jockey for Urmson's job. Things came to a head in May 2011. "Anthony threatens to leave the team if he isn't the single leader," Thrun wrote in an email to Page. "If he is the single leader, a good number of team members will leave."

The situation became so difficult that Thrun was convinced one of them had to go. Late in the spring of 2011 he turned to me for advice on which person he should fire. I told Thrun that both Urmson and Levandowski were valuable members of the team and that he shouldn't fire either of them. Thrun assigned me to deal with the problem, to come in as a kind of grown-up brokering an end to the hostilities between the two bickering engineers. Possibly he picked me because of my experience as GM's vice president of R&D, where I often brokered disputes between brilliant and ambitious researchers. So throughout that summer, on my regular trips to Mountain View, I made time to meet with Urmson and Levandowski to discuss what was going on.

My sympathies were with Urmson in this situation. His Canadian background meshed well with my Midwestern upbringing.

I liked that Urmson was a straight-shooter. I thought he was doing a good job running Chauffeur in difficult circumstances.

Meanwhile, I found it difficult to take Levandowski at his word. There's a great quote about him from a colleague that appeared in a profile Mark Harris wrote in *Wired*—"He's always got a secret plan, and you're not going to know about it." That about sums it up. I've met a lot of smart people in my life, and Anthony is as much of a genius as any I've ever encountered. He'd done some remarkable stuff for Google. But the problem is, he pushes everything to its limits as he pursues his own interests, whether that's the capability of technology or his relationships with other people.

As I worked with Urmson and Levandowski, I suggested to Thrun that he and I get together with the pair of them and spend an entire day talking through the tensions between the two men. In late August 2011, the four of us ended up spending the latter half of a Tuesday together. We had a dinner at a Japanese restaurant, then reconvened the next morning. My approach was to get everything out in the open, to talk through each of the issues that contributed to tensions between the two men. The way they were acting reflected poorly on both of them, I said. On some level, the mere fact that such an intervention was necessary cooled matters. It was sobering—they realized, *this* is what it's come to? Probably my most effective strategy was to remind them of the opportunity they were working toward. Urmson and Levandowski both were brilliant people. At that point, no one on Earth knew more than the two of them about how to make autonomous vehicles. That was the literal truth. Had you, at that moment, ranked all of Earth's 8 billion people by their knowledge of the mechanics of making a driverless car, Urmson and Levandowski would have been tied for first place. I reminded them that they stood to get self-driving cars into the real world much faster working together than apart. Both men realized the truth of my observation.

We left the session with an actionable solution. The result of the brokering was a new governance and execution plan for Chauffeur, which Thrun, Urmson and Levandowski all signed. I insisted on

Chris and Anthony having regular one-on-one discussions so they could continue to air out their differences in private and not let issues spill over onto the Chauffeur team. They needed to build a trusting relationship for Chauffeur to succeed and realize its full potential, and the key to doing this was face-to-face, candid communication. I emphasized they had to function as one team and if they didn't, they would fall short of their aspirations and misspend Google's money. That seemed to settle things between them for a while. But getting to that point involved some awkward moments. Levandowski's strong feelings led him to become quite emotional through the course of one of the conversations.

Meanwhile, Thrun and the Chauffeur team were under increasing pressure to get going on the commercialization of their technology. It was Thrun who made the call on what project Chauffeur should pursue. To him, the greatest potential to get the technology into the greatest number of cars lay with something the team called highway driver's assist. As conceived by Chauffeur, it would be a feature available on new automobiles, which allowed a human operator to drive a vehicle onto a highway, set the car into a suitable lane, then activate the product, which would drive the vehicle for as long as the route required. (A fairly common feature of new cars in 2018, but unknown at the time, and nearly unimagined by most people then.)

— — —

As the Chauffeur engineers debated their future, two other big ideas were setting the stage for a society-wide disruption in the way humans managed their personal mobility. The first was the creation of a car that finally convinced the automotive industry, and the world as a whole, that electric vehicles could appeal to mainstream consumers. The company that achieved the milestone was Tesla.

Tesla has its origins in the partnership between a pair of Silicon Valley entrepreneurs, Martin Eberhard and Marc Tarpenning, who had previously developed an e-book-reader company called NuvoMedia. That start-up sold for $187 million in 2000 to a company called Gemstar, which marketed an interactive television guide. While researching how to power the e-reader, Eberhard and Tarpenning realized that battery-storage capacity had substantially improved, thanks to lithium-ion technology. In fact, Eberhard and Tarpenning realized the technology had improved so much that it might make practical a new type of electric car.

Just as 9/11 played a formidable role in the development of autonomous-vehicle technology, the World Trade Center and Pentagon attacks would also play a role in Tesla's creation. Eberhard and Tarpenning were batting ideas back and forth for new companies in 2001, and after the World Trade Center disaster happened, Eberhard started thinking about the effects of global warming, as well as the distant conflicts that the U.S. fought in the Middle East. Gas-powered cars contributed to each situation, he thought. It just seemed to make sense to come up with an alternative—one that benefited from the new lithium-ion batteries.

A Los Angeles company, AC Propulsion, interested Eberhard with a near-custom battery-electric product line that included midsize passenger sedans and a sports car, called the tZero. Sold as a kit car, the tZero was a fiberglass-body, steel-framed roadster powered by a battery-electric motor, with one version clocking a 0 to 60 mph time of just 4.9 seconds. But the vehicle used lead-acid batteries, and Eberhard knew from his work at NuvoMedia that much lighter lithium-ion batteries provided more power and energy storage. So Eberhard invested $500,000 in AC Propulsion to get them to build him a vehicle powered by lithium-ion batteries. The experience convinced Eberhard to found his own company, with Tarpenning, to build a production version of AC Propulsion's tZero—a nimble, light and ultrafast battery-electric sports car that appealed to the luxury market. The lead investor in the company that Eberhard and Tarpenning incorporated on July 1, 2003, was Elon Musk.

The company would undergo a lot of turmoil before it began shipping its first product, the Tesla Roadster, in 2008. This is not surprising. Anyone who has ever worked in the auto industry will tell you that it's really difficult to mass-produce a new automobile. The 2008 recession didn't help matters. There's also the fact that

Musk, who eventually made himself the company's CEO, could be a difficult personality. At a press event to debut a prototype of the Roadster, an admittedly impressive and fast coupe that could go 0 to 60 in around 4 seconds, Musk said, "Until today, all electric cars have sucked." Which was not only untrue—it was brash and egotistical, a combination that would become familiar to anyone who followed his career.

All the same, under Musk's leadership, Tesla executed some remarkable innovation. It portrayed itself to its employees not just as a company, but as a crusade, as something that could improve the world by finally reforming the system of automobile transportation. It may have been the first company to pit Silicon Valley against Detroit. It was definitely the first to sell a mass-produced battery-electric vehicle that used lithium-ion batteries. Tesla went public in the summer of 2010, with the first IPO of an American car company since Ford did it in 1956. It also represented the first creation of a major new car company since the founding of Chrysler in 1925.

When Tesla rolled out the Model S in 2012, the vehicle redefined the way the marketplace perceived electric cars. GM's EV1 was peppy and fun to drive, but its limited production numbers prevented it from making a big impact on consumer attitudes regarding what an electric car could be. The Toyota Prius hybrid was a fine car, but its appeal derived from its small environmental footprint rather than a cool factor that appealed to car guys. The Tesla Model S was the first mass-produced alternative-propulsion vehicle to succeed on its own terms. It was a beautiful car that handled well, accelerated like a rocket and just happened to be battery-electric. I, of course, liked that the Model S extended the skateboard-chassis idea pioneered by my team at General Motors with the Autonomy concept. With a 0 to 60 mph time of 4.2 seconds, the four-door luxury sedan had a single-charge range in the neighborhood of 250 miles. Some versions could seat up to seven people. Factor in great safety ratings, an enormous touch-screen console and a series of neat little touches, like door handles that appeared only when the driver approached the vehicle, and the Model S wasn't just a cool electric car—it was a cool car, period. That year, the Model S was named *Motor Trend*'s Car of the Year—the first vehicle to win the award that wasn't powered by gas.

—

The second big idea setting the stage for a society-wide disruption in the way humans managed their personal mobility began with a Boston entrepreneur named Robin Chase, who cofounded Zipcar in 1999.

The start-up called itself a car-sharing firm. For a nominal membership fee, it provided its urban members with access to a nearby vehicle for a reasonable, hourly rate. Really, it was a short-term form of car rental, one that was innovative on a number of levels. It distributed its selection of hip urban models in parking lots located close to where its customers lived, and made them available by the hour as well as by the day. Instead of going to a car-rental location, a customer had only to walk up to a Zipcar and use a credit-card-size badge to open the vehicle. The keys were hidden inside, and the customer could head off wherever they chose: the grocery store or big-box retailer or on a weekend road trip.

Chase believed she was breaking the car-rental model, and combatting the hegemony of the idea that every American adult had to own at least one car. "In the early days, I had this *Godfather* image in my head, these guys bursting in with machine guns and taking me out," she said. Actually, the people in the established rental car companies scoffed at Zipcar. Detroit dismissed her as well. "Robin, you don't understand," people would tell Chase. "People's self-esteem is tied to their cars."

Certainly, that *used* to be true. When I was growing up in the fifties and sixties, car ownership seemed mandatory, and not only that—you had to own the right *kind* of car. But Zipcar's rise reflected and extended the change in consumer attitudes toward automobiles, as did an increasingly urban population. Social media also played a role, as well as the rise of smartphones. Today's youth don't regard car ownership as mandatory. In 2000, sixteen- to thirty-four-year-olds purchased 5 light vehicles per hundred people. In 2015, the statistic had fallen to 3.5 light vehicles. Over the same period, the average age of new vehicle buyers increased by nearly seven years, according to a 2016 article by Federal Reserve economists.

What's changed, says Chase, is where kids want to put their first discretionary dollars. They get a part-time job and they don't save up their money to buy a car. Now they're saving up for a new cell phone or maybe a video game console. "They would rather have a thinner iPad than a cool used car," Chase observes.

Chase was confident in the car-sharing model despite the criticism because of her belief that "economics and convenience trumps status"—that a new product will succeed if it provides the same service as an existing product with greater convenience and at a dramatically lower price. "Zipcar was more convenient than having your own car," she said.

It also was among the factors that inspired the next transportation innovator in this story. Logan Green grew up in the Los Angeles area. In high school, a part-time job at Atari founder Nolan Bushnell's gaming company, uWink, required Green to commute on the Los Angeles freeways. Green despised the traffic caused, he figured, by the three-quarters of car seats that went unfilled on most trips. "I just recall having this feeling of seeing everyone stuck in traffic," Green told author Brad Stone in an interview for Stone's book *The Upstarts*. "There were thousands of people heading in the same direction, one person in each car. I thought, 'If we can just get two people in the car, you could get half these cars off the road.'"

In college at the University of California, Santa Barbara, Green heard about Zipcar and attempted in 2002 to get the car-sharing service to deploy a fleet at the school. Chase had only one hundred cars at the time and declined, but the experience inspired Green to set up something similar to Zipcar on his own. UCSB bought a fleet of Toyota Priuses, and Green whipped up a Zipcar-like tech infrastructure, including an online reservation system and the capability to unlock the cars with access codes and RFID (radio-frequency identification) cards. The system tended to be used for short, intra-city trips. To return home to the L.A. area, Green relied on bus lines or the ride-share section of the local craigslist website.

During a summer of traveling in Africa, Green was struck by the casual way the citizens of Zimbabwe hired rides. No one used actual, licensed cabbies. Instead, they just reached out an arm and hailed the nearest car with an open seat—and expected to kick in some gas money as a fare. Logan realized that Zimbabwe, a developing country, may have had a more efficient transportation network than Santa Barbara. He should know—by that point, he'd become the youngest member of the Santa Barbara Metropolitan Transit District board.

In his last year at Santa Barbara, Green came up with a concept that melded together all of his experiences in an attempt to increase the efficiency of the California transportation system. He founded a kind of online hitchhiking service, which he dubbed Zimride, for "Zimbabwe ride." Green's start-up used the Internet to pair drivers and their empty seats to the passengers who required a lift to the same destination. Around the same time, Facebook opened its service to other software developers, and Green jumped on the opportunity to use social media to connect drivers and riders in a way that also made it easy to investigate the profiles of the person providing, or sharing, your ride. Zimride's first app, known as Carpool, used Facebook to pair university-student drivers with other college kids who were looking for rides. The service spread from Santa Barbara with help from Facebook, which touted the app as an example of the way programmers were using the platform to provide new and innovative services.

Recent Cornell graduate John Zimmer noticed Green's company in part because of the coincidental similarity to his own name. He joined Zimride on a part-time basis while he worked as a real-estate analyst for Lehman Brothers, and for some time the start-up ambled along as Green and Zimmer's side hustle, functioning as an online ride-board for colleges across the country, as well as the occasional corporation. Green and Zimmer opened it up to everyone else in 2010 and launched regular minibus routes between places like San

Francisco and Los Angeles. But the idea didn't catch fire. So in the spring of 2012, the Zimride leaders and their programming team began discussing other ideas to solve the transportation problem.

One of the things that excited the Zimride guys was a service called Uber. The company was first registered with the state of California as UberCab on November 17, 2008, by Garrett Camp, a Canadian entrepreneur. While working on his master's degree at the University of Calgary, Camp had cofounded StumbleUpon, a first-wave social media site designed to help people recommend interesting things they found on the web. He sold the company to eBay for $75 million. By that point he'd moved to San Francisco. He bought a Mercedes-Benz sports car, but on nights out he preferred to leave that home and use cabs—and quickly grew frustrated with San Francisco's notoriously terrible taxi service. So Camp transitioned to the city's gypsy cabs, usually late-model black sedans that solicited passengers with a flash of the headlights. That experience, along with Camp's fascination with a scene in the James Bond movie *Casino Royale,* in which 007 used a smartphone to track an automobile with a mobile map, prompted a brainwave. Camp realized that he could use the iPhone and its sensors to make like James Bond, and hail San Francisco cabbies with a software program downloaded from Apple's just-launched App Store.

Shortly after, in December 2008, Camp traveled to Paris to attend the LeWeb tech conference with Travis Kalanick, an old friend. Kalanick, a UCLA computer science grad, was a serial entrepreneur who had been an early employee of the file-sharing network Scour, a Napster competitor that also allowed the trading of video. Searching for his next big thing at LeWeb, Kalanick's main idea was for something very like Airbnb. Camp was trying to convince Kalanick to work on his own idea, the already registered UberCab, when Camp and Kalanick took a taxi ride with Camp's on-again, off-again girlfriend. During the ride, Camp's girlfriend placed her feet, in high heels, on the cab seat. When the driver yelled at them, they left the cab in a fit of anger.

That was the turning point for Kalanick, who decided to try to improve the experience of hiring a cab, and automobile transportation overall. By the time the three of them returned to San Francisco, he'd decided to join in on Camp's idea. Kalanick and Camp hired Ryan Graves as their first full-time employee in January 2010. UberCab went live on Apple's App Store in June 2010 as a tool that allowed people in San Francisco to summon one of the city's black car limos. "UberCab is everyone's private driver," read an email that went out to potential investors shortly after. A battle with the San Francisco taxi regulators brought publicity to the start-up, spurring growth in users of 30 percent a month. The buzz convinced Kalanick to join more formally as a CEO. And in 2011, the company expanded its services to New York City.

Within just a couple of years, two companies, Uber and Lyft, would be firmly established as the yin and the yang, the Coke and the Pepsi, of the ride-sharing world. The pair do have a lot of contrasts. Kalanick's prickly personality is legendary—"It's hard to be a disrupter and not be an asshole," explains one of Uber's earliest investors in a *Vanity Fair* profile of Kalanick. In contrast, early investors of their company worried that Green and Zimmer were too *nice* to be entrepreneurs. Lyft began its rides with a fist bump between driver and rider, encouraged early drivers to stick a pink mustache on their front grill and played up the social aspects of mutually beneficial transportation. Meanwhile, Uber's minimalist design ethos encouraged people to grasp the efficiency provided by the new way to arrange transportation. Whatever their differences, the two competitors would usher in ride sharing to the world.

Chapter Nine

THE $4 TRILLION DISRUPTION

An optimist will tell you the glass is half full;
a pessimist, half empty; and the engineer will tell you
the glass is twice the size it needs to be.

—UNKNOWN

During the summer and fall of 2011, Anthony Levandowski often asked me how much I thought the market for autonomous vehicles was worth. At the time, I didn't know anything about the Chauffeur Bonus Plan. I didn't know that Levandowski had the largest share, nor did I know that it was tied to the valuation of the self-driving-car project once it was spun out of Google. Separate from all that, however, I could provide Levandowski with some information that would go a long way to satisfying his curiosity, because I was leading a research project into the effects of the four disruptive forces that had been unleashed on the American transportation system.

DARPA, Carnegie Mellon, Stanford and Google had demonstrated that autonomous vehicles were viable. GM's EN-V concept, a star of the Shanghai World Expo, had highlighted the promise of vehicles tailor-designed for our most frequent trips instead of being over-designed to handle all conceivable trips. Zipcar, Lyft and Uber were divorcing Americans from the idea that you had to own the vehicle you rode in. And Tesla was making electric vehicles mainstream. Individually, each of the trends offered significant improvement over the 130-year-old automobile transportation system. But

I was more interested in what could result from combining them. I sensed a new age of automobility was imminent. I believed this new age promised better mobility and safety for more people at lower personal and environmental cost, and I wanted to more deeply understand the impacts of the convergence. Just what sort of a future were these trends going to unveil? And to speak to a point that was relevant to the Google engineers who were included in the Chauffeur Bonus Plan—just how much, in economic terms, was the market for disrupted mobility shaping up to be?

Jeff Sachs provided me with the opportunity to gain this understanding. Jeff is one of the world's handful of celebrity economists—a longtime director of Columbia University's Earth Institute, which sought novel ways to increase the sustainability of the human lifestyle, and the author of the 2005 bestseller *The End of Poverty*. I knew Jeff's reputation and really started paying attention to him in November 2008, when he wrote an op-ed that argued the counterintuitive idea that the deepening financial crisis represented an *opportunity* for Detroit to begin a new era of U.S. technological leadership in the global auto industry. I agreed with Jeff's perspective.

After Obama's election, Sachs advised the president on the auto crisis and came to Detroit to learn more about GM's situation. I spent an afternoon with him in early 2009 discussing the potential of the new DNA of the automobile. It was clear we shared a common vision for the future of transportation and a passion to realize this vision as soon as possible

Sachs contacted me after I left GM in the fall of 2009. He wanted me to head up a new initiative at his institute—something he proposed calling the Program on Sustainable Mobility.

I liked Sachs a lot and agreed to join the Earth Institute in January 2010.

— — —

With an opportunity to conduct research into sustainable mobility at Columbia's Earth Institute, and unfettered by any bonds with General Motors, I felt liberated to investigate the impact of autonomous, shared, electric and tailored vehicles on America's transportation costs. To launch the research, Jeff helped me scrape together enough funding from six companies. (The companies were auto manufacturers General Motors and Volvo, telecommunications firms Ericsson and Verizon, the electrical utility Florida Power & Light, and the real estate developer Kitson & Partners.) I hired an outstanding program manager, Bonnie Scarborough, an engineer who had previously worked at the National Academy of Engineering, as well as a couple of research assistants. I also convinced my good friend and colleague Bill Jordan to join the team.

Bill is one of the best math modelers in the world. I recruited him to GM R&D in the early eighties after he completed his PhD in civil engineering at Cornell. We quickly became good friends and collaborated on many challenging and interesting projects that applied our math and statistics skills to improve GM operations and products. Bill left GM the same day I did and remained active professionally as a consultant. When I told him what I was up to at Columbia and why I needed his help, he signed on immediately.

To begin the work, we met regularly in the backyard of my house in Franklin Village, just northwest of Detroit. One of the first things we did was go through a series of calculations that estimated the cost to Americans of their automobile driving. As a nation, our citizens drive more than 3 trillion miles per year—and the cost of this travel is enormous. The American Automobile Association (AAA) annually estimates the out-of-pocket cost of owning and operating automobiles. Their calculation includes vehicle depreciation, fuel, insurance, maintenance and finance. Some of these costs are incurred with each mile driven (e.g., fuel and depreciation) while others are incurred annually (e.g., insurance and finance).

AAA's 2011 estimate for the out-of-pocket cost of a typical car was about $0.60 per mile, not including parking. When you add in parking costs per mile of about $0.05 (which varies widely with where you live and where you park), the out-of-pocket cost ends up being around $0.65 per mile. This means Americans spend $2 trillion per year in cash to own and operate their vehicles (because we drive 3 trillion miles a year times $0.65 per mile).

We also needed to account for the cost of the *time* we spend driving cars. While economists have worked diligently to estimate the value of travel time, this cost remains highly contentious. Bill and I calculated that by simply dividing the average American worker's wage per hour by the average distance driven per hour. In 2011, the average U.S. wage was $24 per hour ($43,000 earned per year divided by 1,800 hours worked per year) and average driving speeds in U.S. cities ranged from 25 mph to 30 mph (this average includes the time you spend stuck in traffic and stopping for red lights). This result is a time cost of around $0.85 per mile ($24 per hour divided by 28 mph). Add the typical out-of-pocket cost of $0.65 per mile to the time cost, and the total cost of owning and operating a car is around $1.50 per mile.

Altogether, then, Americans spend about $4.5 trillion per year driving their automobiles (3 trillion miles per year times $1.50 per mile). This is an enormous amount of money—more, in fact, than the nearly $4 trillion annual budget of the U.S. federal government! Bill and I were convinced the $4.5 trillion being spent by Americans each year to own and operate their cars would be reduced substantially by the mobility disruption.

Here's how we saw the new system working: A person who needed to get somewhere—say, the grocery store a few miles away—would use a smartphone to summon a driverless, electric vehicle. This request would be managed by a dispatch computer, which would assign the ride to a specific vehicle drawn from a shared fleet. The vehicle would drive itself to the specified pickup point. The passenger would climb into the car, and car and passenger would head to the grocery store, where the passenger would get out. The car would then head autonomously to pick up its next rider. Or head to a staging area, where it could be cleaned and fueled. Then it would wait for the next call.

How long was a reasonable time for someone to wait for a ride? How many fleet vehicles would be required to meet this response time requirement? How many miles loaded and empty would each vehicle travel each day? What was the best way to assign vehicles to customers? These were the sorts of questions Bill and I needed to answer to estimate the size of the market for this type of service.

Our work would require some complex math and analysis, but Bill and I were comfortable with that. This was the sort of challenge that we'd tackled together often at GM. We loved solving such problems.

Compared to a personally owned and operated vehicle, which sits motionless and empty most of the time, our service would utilize vehicles at a much higher rate. Rather than being parked, they would be moving people around. But to do this, they needed to travel empty to pick up their next customer. Lots of empty miles traveled by lots of driverless taxis could add up to a significant expense. Would the expense of the empty miles negate the savings in fleet utilization and parking? How big would the fleet have to be to ensure a vehicle is nearby when it's needed? Our math model needed to find the sweet spot between customer response times, fleet utilization and empty miles. And we had to account for the randomness of trip start and end points and morning and afternoon rush hours. It would not be sufficient to get to customers quickly on *average*. We also needed to be sure we avoided the occasional really long wait that would upset any rider.

We began by pulling together data. The length of a typical car trip. The average trip speed. The amount of time needed per trip. Of course, such figures change depending on which town or city you're looking into. We needed to select a specific location to make our calculations relevant to real people. Ultimately, we modeled a number of different locations selected to be representative of the various environments in which Americans lived. The first was Ann Arbor, because it seemed a typically American small city, much like similar communities across the country. Plus, it had the added advantage of being located just a short distance away from where Bill and I lived in Michigan. We also created calculations for Manhattan, as the densest urban setting in the United States. What sort of cost savings, we wondered, would the mobility disruption provide to those who lived in such settings?

———

Ann Arbor, Michigan, has a population of 285,000 people and is home to the University of Michigan. In addition to my Columbia University position, I had signed on as a professor of engineering practice at U of M in the spring of 2010 and was familiar with the area. Coincidentally, it was the city that inspired Google cofounder Larry Page's dreams of disrupting transportation.

I felt many of the same frustrations with Ann Arbor's transportation system that Page did. My office was located on the North Campus, and I often gave lectures on the Central Campus, about 2.5 miles away. I typically drove to the Central Campus so I could head home in my car following my lectures. The biggest challenge was finding a place to park. U of M staff could purchase a special permit for $800 a year that allowed one to park in select structures and lots throughout Ann Arbor. Even though I had one of these permits, to ensure I arrived on time, I typically left my office 45 minutes before class—allowing 10 minutes to drive the 2.5 miles, 20 minutes to find a parking spot (often driving up eight levels in a parking deck with my fingers crossed) and then 10 minutes hustling on foot to the lecture hall. Such experiences only increased my motivation to disrupt such inefficient mobility systems.

To analyze the impact of an autonomous mobility-on-demand service on Ann Arbor, Bill and I made a series of assumptions. We set out to model a world where this service matched or exceeded the convenience of personal car ownership. So, how soon would one of these shared, autonomous vehicles have to arrive after being summoned? About the same amount of time required to find your keys, get to your garage, start your car and back out of your driveway. Convenience, we decided, would amount to an autonomous car arriving within *two minutes* from when it was requested—a fairly stringent standard designed to lead to conservative cost estimates.

We then assumed the fleet operator knows the current location of all its vehicles and can estimate when they will complete their trips—as is currently the case with such companies as Uber and Lyft. We also assumed the operator can estimate when each vehicle could potentially reach the new customer's location. We further assumed that the operator would keep its customers informed about assigned vehicle locations and expected arrival times. (Uber and Lyft have been providing all this information to its customers for several years, and by now some of us take them for granted. But when we were conceiving this model in 2011, these notions seemed revolutionary—and really, they were.)

We made other, more arbitrary assumptions to simplify the math: that the region served by the vehicle fleet was a square. And that the trip origins and destinations were evenly spread out—an assumption that certainly isn't reflective of many cities in America, where traffic tends to flow from the suburbs into a downtown in the mornings, with the reverse being true in the evenings. We also assumed that vehicles are randomly scattered throughout the service region, which was a simplification, since there'd likely be numerous parking and maintenance facilities throughout the area. These are the kinds of assumptions math modelers make to gain initial insights. We knew we could relax them later using sensitivity analysis and simulations.

Bill built the mathematical models, then ran basic simulations on a hypothetical urban area (not Ann Arbor—just a general urban area). At this point we were simply testing the software. Even with these preliminary models we were astonished by the results that came back. "That can't be right," I told Bill at the table in my backyard. "We have to run the numbers again." The answer our model provided was just as astonishing the second time around. Same with the third. In each run, after double- and triple-checking our work, the results said the autonomous shared mobility system we envisioned would be able to respond quickly to customer requests, with less than two minutes response time on average. The distance the vehicles would travel while empty was low, at just about 5 percent of loaded miles. That meant fleet utilization was high, with vehicles serving customers about 75 percent of the time from 6:00 A.M. to 8:00 P.M. More amazingly, this outstanding performance could be attained with a fleet size equal in number to just 15 percent of the population being served.

Okay, maybe we were on to something, I thought. But what if we used real-world data for Ann Arbor in the models? How many shared vehicles were required to ensure adequate wait times of about two minutes per trip request? According to the federal government's 2009 transport survey, Ann Arbor had 200,000 personally owned vehicles that made 740,000 trips a day. Between 6:00 A.M. and 8:00 P.M., the vehicles had a utilization rate of about 8 percent, meaning they were in use for an average of just 67 minutes a day.

We focused on the 120,000 vehicles that were driven less than 70 miles a day—figuring that people would be most likely to use a shared, driverless fleet for the trips that were taken *within* the Ann Arbor urban area. These vehicles were responsible for 528,000 trips a day, for an average of 4.4 trips per vehicle per day, carrying 1.4 occupants an average distance of 5.8 miles.

To calculate the size of the fleet required to serve all these trips within Ann Arbor, we used math models developed for queuing systems, that is, systems where people or things wait in line for service. (Highway intersections and grocery checkouts are examples of queuing systems.) We found that an astonishingly low number of vehicles could provide customers with average wait times of less than a minute. Ann Arbor required just 13,000 vehicles to service the city's internal trips at average periods. To provide nearly instantaneous access for trips in and around Ann Arbor, even during rush hours, the city would require a shared fleet of just 18,000 vehicles.

When the computer spit out that result, I was floored. I couldn't believe it. The figures matched our math models. An 18,000-vehicle fleet was just 15 percent of the total vehicles being used for transport in and around Ann Arbor.

How could Ann Arbor's mobility needs be serviced by such a small fleet? The answer was tied to population density and the number of trips people make each day. It turned out to be highly likely that someone would request a ride within minutes of someone else very nearby completing a ride. This meant the empty vehicle needed to move just a short distance autonomously and could get to a customer quickly. And by buffering the fleet size by 5,000 vehicles to accommodate rush hours and other surges in demand, about 40 vehicles per square mile were typically unassigned at any given time. Because they were autonomous, these cars could drive themselves to strategic locations around the city so they could reach customers quickly.

But was Ann Arbor unique? We decided to study several other cities: Salt Lake City, Utah; Rochester, New York; Columbus, Ohio; Austin, Texas; and Sacramento, California. In every case, a shared fleet size equal to about 15 percent of the number of cars owned in these cities would provide responsive and efficient mobility service. Upon further analysis, we concluded that in communities with population densities in excess of around 750 people per square mile, our envisioned mobility system based on autonomous shared vehicles would perform quite well. This density criterion covers most towns and cities in the U.S. In addition, we were encouraged by how robust our results were to changes in key modeling assumptions. For example, you'd need to serve only 10 percent of the trips in an urban area to simultaneously attain high fleet utilization, low empty miles and fast response times. This means you could launch a commercial business with a relatively modest market share of trips.

— — —

Thanks to our calculations, we were beginning to grasp how revolutionary this approaching mobility disruption could turn out to be. But we still needed to estimate the cost of shared, electric and autonomous mobility compared to the $1.50 per mile Americans typically incurred to own and operate a car. Would the cost of the new system be prohibitively high because the technology was expensive? Or could subscribing or hiring a driverless vehicle reduce transportation costs significantly?

Because 75 to 85 percent of American car trips are one- or two-person, we decided to model the cost of a tailor-designed two-person vehicle like the EN-V. While we worked at GM, Chris Borroni-Bird and I had estimated EN-V would weigh less than 1,000 pounds, one-third to one-fourth what a typical vehicle weighs, and have 90 percent fewer parts than a normal car. We had concluded it could be built for around $7,500, not counting the self-driving system. So, to be conservative, Bill and I assumed our vehicle would cost around $10,000, excluding the autonomous-driving technology. Based on 250,000 miles of use, on par with a taxicab, depreciation cost amounted to $0.04 a mile. The cost of electricity to charge the batteries amounted to about $0.01 per mile. (In contrast, an efficient car in 2011 getting 30 miles per gallon had a $0.10 cost per mile for fuel at $3.00 per gallon of gasoline.) We set maintenance costs at $0.05 a mile, about what AAA said applied to a conventional car. (Although electric cars require less maintenance, we wanted to allow enough money to replace the batteries.) Insurance costs could be cut significantly compared with that of conventional cars, because traffic safety experts predict that driverless cars could eliminate at least 90 percent of car crashes. So rather than the $0.05 to $0.10 a mile that insuring conventional cars required, we assumed $0.02 per mile for insurance. Parking costs would be negligible because the fleet was highly utilized so we reduced them from $0.05 to $0.01 per mile. And we added $0.01 per mile as the cost of financing.

When we added up our estimated costs, we hit $0.14 per mile, which we then boosted by 10 percent to account for the cost of the empty miles traveled between riders. So, we concluded, our shared, tailored electric vehicle would cost around $0.15 per mile before adding in the cost of the driverless system.

Now to calculate the cost of the technology. Driverless systems required lasers, radars, cameras, computers, digital maps, software, actuators—which in 2011 existed only as prohibitively expensive prototypes. But new technology is always expensive in the prototype phase. As engineers with long careers in the auto industry, Bill and I expected the material costs to decrease as the technology matured. Anti-lock brakes, electronic stability control, hybrid propulsion systems, advanced batteries and fuel cells—the costs of these technologies have dropped by factors of five to ten as they matured. Engineers are really good at improving products once they know they are viable and offer value.

Bill and I decided to assume the driverless technology for our envisioned mobility system would cost $10,000 per vehicle. (This turns out to have been conservative; recent forecasts by consulting firms put the figure around $5,000.) The additional $10,000 added another $0.04 per mile ($10,000 divided by 250,000 miles of use). We then threw in another $0.01 per mile to account for financing this cost and the electricity required to power the on-board processor and sensors.

Adding the $0.05 per mile for the autonomous system to the $0.15 per mile for the vehicle, we concluded a tailor-designed, electrically driven, two-person, shared, autonomous vehicle would cost on the order of $0.20 per loaded vehicle mile. Compared to our $1.50 estimate of the cost of owning and operating a car, our envisioned mobility system would decrease the expense of convenient transportation by a whopping $1.30 per mile.

We were excited. To our knowledge, this was the first time anyone had estimated the combined potential of autonomous, shared, electric and tailor-designed vehicles. Multiplying the $1.30 per mile savings by the 3 trillion miles Americans drive annually revealed how much the mobility disruption stood to save drivers in our country. The new age of automobility could reduce America's $4.5 trillion per year mobility bill by $3.9 trillion per year. Even if our cost estimates were off by a factor of two, which we thought unlikely, the savings would remain in the trillions. All told, our calculations suggested that employing a shared driverless and electric mobility solution could save a driver $5,625 a year, plus the value of all the time that not driving frees up. Depending on how you value the time, the annual savings could amount to $16,000 a year, or even higher.

The analysis we performed for Manhattan was just as exciting for those looking to gain the benefits of much cheaper transportation. Home to 1.6 million people living in just twenty-three square miles, the New York borough of Manhattan sees relatively low rates of car ownership because of high parking costs, the easy availability of relatively reasonably priced taxis and an extensive public transit system. So how would a shared, driverless fleet compete with Manhattan's famed corps of cab drivers? To calculate that, Bill and I started crunching the numbers. Back in 2011, about 410,000 taxi rides per day happened internally from one point to another in Manhattan. The average wait time for one of those passengers was five minutes, and an average ride cost the rider about $5 per mile. We calculated that it would take about 9,000 driverless shared vehicles to service those 410,000 trips per day within Manhattan with an average wait time of under a minute—much less than the average 5 minutes that it took to get a cab. The price would be a lot more reasonable, too. Taking into account a 15 percent profit margin for the fleet operator, we estimated that the price of near-instant access to a shared, driverless vehicle fleet could be about $0.50 a mile, or about a tenth of what the customer pays to hire a Yellow Cab. Putting it slightly differently, the average cab ride within Manhattan costs the consumer $8 plus tip. Meanwhile, Bill and I concluded that a shared driverless fleet could provide a better ride experience at just $1 a trip.

For the first time, I grasped how big this thing was. The profit and potential shareholder value were enormous. So was the threat to the incumbent players in the century-old automobile transportation system. And the disruption seemed inevitable. As Robin Chase would observe years later in a prescient 2016 piece in *Wired*'s Backchannel blog, "While one city, or one state, or one country might try to slow it down, there are many others that will step forward to lead the way. No matter how protracted the fight and the transition, we are going to end up choosing self-driving cars . . ." The rapid growth of Uber and Lyft showed that individuals were willing to change the way they get around to save a little money. Eliminating the driver from the equation, and the material and energy waste from the system, stands to save people a heck of a lot more money than those early, human-driven iterations of Uber and Lyft—and as a result, the new system will be adopted that much more quickly.

The new age of automobility promises to profoundly improve the way regular people live their daily lives. It democratizes the benefits of convenient, whenever-you-want, wherever-you-want transportation to people who couldn't previously afford it: the very young and the very old, who aren't able to drive because they don't yet or no longer have licenses. The disabled, who somehow lack the ability to operate a motor vehicle. Everyone else would enjoy the benefits of more reasonably priced mobility.

To get an idea of the way the new mobility system would work, consider a morning in the life of a typical family—let's call them the Wilkersons, and assume they live in the Chicago suburb of Evanston, Illinois. Thirty years after the 9/11 attacks on America that kick-started everything, the morning of September 11, 2031, happens much like most in the Wilkerson home.

Over a half-eaten bowl of cereal and some toast, nine-year-old Tommy Jr. plays a virtual-reality computer game while his eleven-year-old sister, Tammy, texts friends about the day ahead. On the other side of the breakfast table, parents Mary and Thomas are swiping through the day's news on their holographic displays when Tammy looks over at her brother. "Tommy, did you unload the goods closet yet?"

Thomas nods. "You're supposed to do that before breakfast every day," he reminds his son. "I'll help you this morning."

Together, father and son troop from the kitchen to the recreation area that used to be the home's two-car garage. (Like most families, who no longer own their own vehicles, the Wilkersons have converted the garage to living space.)

The goods closet is on the far side of the rec area. It's a shelf-filled cupboard whose air-lock-like design can be opened from both the inside and the outside of the house. Each night, delivery drones drop off things like food orders, home-office supplies and other items the family has ordered online. This morning, in addition to the usual groceries, like milk and salad greens, there's a long cardboard box emblazoned with the logo of a company called Hyperlite. "My new wakeboard!" Tommy shouts.

"Not till we get to the cottage," Thomas says, placing a hand on his boy's shoulder.

Soon after they've put away the groceries, a smartphone buzz alerts the Wilkersons that their ride is arriving—the result of a standing arrangement Mary has with their car-sharing company, Maghicle, to which they pay a monthly subscription fee in exchange for on-demand use of a company vehicle for a certain number of miles per month.

"Bye, Grandma!" shouts Tommy to his father's mom, who will be picked up shortly by a two-person mobility pod that will take her to a bridge tournament—one of numerous recurring social engagements the eighty-five-year-old woman is able to maintain in a world of disrupted mobility.

Outside, the sky is clear. Smog warnings and acid rain are a forgotten element of a bygone era because nearly all vehicles are powered by batteries or hydrogen fuel cells. Little Tommy clambers into the four-seat self-driving car that waits for them at the curb. Next comes Tammy, and Mary, smartly attired for her job as a partner at the PR firm she cofounded. Last is Thomas, dressed in a suit for an important meeting. Years before, Thomas's parents owned an auto dealership. Then Thomas pivoted that company into one that operates staging areas for driverless vehicles. The Wilkersons' depots are located around the Chicagoland area and provide charging, cleaning and maintenance services to some of the nation's largest operators of self-driving fleets. In fact, today Thomas is meeting with an investment bank in downtown Chicago to discuss a deal to purchase the company's thirty-eighth staging location.

"Can I say it today, Mom?" Tommy asks, and at his mother's nod, the boy clears his throat. "Maghicle," he says in a serious tone. "Ride begin."

The doors lock, and almost imperceptibly, the automobile starts to move. Years ago, the jerky start-stop motion that was inevitable when humans operated automobiles meant many passengers became too motion sick to read or stream video in their vehicles. But driverless cars are designed to make acceleration and deceleration imperceptible. Some pods have seats and desks designed to allow people to work throughout the rides; others feature more comfortable recliners conducive to virtual-reality experiences.

Decades before, morning family commutes were hectic as parents used their own wits and driving skills to pilot family vehicles through rush-hour traffic to arrive before the morning school bell. But in a future of disrupted mobility, the school commute is pleasant and relaxing—an opportunity for parents to spend quality time with their children. During the two-mile ride to the children's school near the west side of Evanston, Tammy and her father pass the time chatting about Tammy's upcoming tryout for the school's volleyball team. Meanwhile, Mary marks Tommy's homework and applauds him for getting nine out of ten on his math review.

Once the kids are dropped off, the vehicle's route-planning software calculates the most efficient path downtown and sets off to transport Thomas and Mary via the southbound lanes of the I-94 freeway. Taking a break from reading his term sheet, Thomas looks out the window to keep an eye out for potential new locations for the family business. Beside him, Mary keys through a deck she'll tele-present later that day for a client in Shanghai. How different this commute is from the one his father took, decades before, Thomas thinks. Maghicle's surge-pricing model cooperates with similar models maintained by its competitors to encourage a more even distribution of traffic. The highest surge pricing is happening now, between 8:00 and 9:00 A.M., and while Thomas and Mary don't love paying the extra fees, they do it because they like to accompany their kids to school.

On the highway around Thomas and Mary are a variety of driverless vehicles, from large tractor-trailers to smaller two-person "pods." Even during rush hour, the commute happens smoothly and quickly. Complex algorithms manage the interplay of the autonomous vehicles, eliminating crashes, maintaining a safe distance between them at all times and orchestrating merging at on- and off-ramps. The commute from Evanston that once took Thomas's father an hour now lasts just thirty minutes for Thomas and Mary.

Thinking about his father reminds Thomas of the family cottage, and with a few swipes in the air on his free-floating screen, he reserves a larger hybrid sport-utility vehicle for the next day's trip to their vacation home on the opposite side of Lake Michigan.

Minutes later, Thomas and Mary's vehicle gets off I-90 and slows to a stop at the Loop intersection of Madison and LaSalle, which looks a lot different from the way it did thirty years before. The streetscape is designed now for pedestrians rather than automobiles. The valuable space once devoted to on-street parking has given way to wider and greener sidewalks. Former parking lots have been converted to parks, cafés or public plazas. Other lots now host residential condominiums or new office buildings.

Thomas and Mary exit their vehicle. The vehicle whizzes off on its own, headed either to pick up its next rider or to a nearby staging area to wait for its next assignment. On the sidewalk in the Loop, the couple give each other a quick kiss goodbye.

"See you here at five," Thomas says, and they head off to begin their separate days, in a world that is both different and similar to the one where we live today.

To get to the Wilkersons' world would require a massive disruption in numerous different businesses—and the calculations performed by Bill Jordan and me suggested this disruption was not only likely but inevitable. Bill, Bonnie Scarborough and I shared our results with our Earth Institute sponsors late in the summer of 2011. Jeff Sachs and I then tried to convince our sponsors to pursue the opportunity as a business. But even though some of them, such as GM and Volvo, should have realized how close these technologies were to being realized, no one bit. Evidently, it all sounded like science fiction to them.

So I started talking about the opportunity to pretty much anyone who would listen. In casual conversation with Google staffers like Chris Urmson and Anthony Levandowski, among others, I shared the broad details of the research results. It fascinated them on a number of levels. Intrigued, Urmson asked me to present the results of my research to the greater Chauffeur team.

Urmson and I set a date in December 2011 for me to make that presentation. Before that date arrived, I presented the findings at the annual gathering of the International Federation of Automotive Engineering Societies, in Mainz, Germany, in November. Known as FISITA, the meeting was one of the key ways that automotive engineers and executives communicated to one another what was happening in the industry.

"We've entered into a very important period of time for automotive engineering and design," I told the group at the beginning of my speech, mentioning alternative-propulsion systems, tailored designs and "cars that can literally drive themselves." I showed a little movie of Boss at the DARPA Urban Challenge and mentioned Chauffeur's efforts in autonomous driving. Google's self-driving cars, I said, "can stop at a traffic light and can detect and avoid a pedestrian with a baby stroller.

"I advise Google," I said. "I ride in these cars. Month by month you can just see the learning curves that all of us as engineers can appreciate . . . We've gone from laboratories and experiments and proving grounds and controlled competitions to actual public road experience . . . The progress that's being made is dramatic."

Then I took the audience through the journey that Bill and I had taken with our research. Thanks to this new mobility system, I said, many cities could manage their intra-urban transportation with only about 15 percent of the vehicles they had now. These shared fleets would provide the same convenience of use and freedom of mobility as conventional, personally owned automobiles. Would they save any money for the users? In fact, they would, I told the audience. "Our analysis of Ann Arbor and several other U.S. cities indicate we could reduce the cost of mobility from $1.50 per mile to about $0.20 per mile.

"So, is this a fairy tale or not?" I asked the crowd of engineers. "This is a huge opportunity through the eyes of the consumers. What I'm telling you here today is that you can take a community like Ann Arbor and reduce the cost of mobility by over eighty percent—and still have the same spontaneity and travel patterns that you would have by owning your own vehicle."

I wrapped up the speech by describing my belief that the automobile industry was about to be disrupted by technology, same as the photography, media and music industries had. "Incumbents rarely do well when industries disrupt," I said. "There's something about being a big player in a mature industry that redirects all of the attention to the next quarter's profit . . . I would contend that the transportation and energy industries are as ripe for disruption now as all of these industries have been.

"The fact is, it's within our grasp, and I believe the consumers are going to love what we come up with—and I think there's going to be ways to make some really good money doing it. So," I finished. "Why are we waiting?"

I suspect that there were engineers and executives in the audience from every major automaker. So far as I know, not a single one of them rushed from Mainz back to their corporate headquarters and raised the alarm—that technology *currently available* could converge to provide transportation at less than a fifth of its current cost. Not only that—the same technology could radically decrease the number of lives lost to automobile crashes. What a missed opportunity.

— — —

In fact, one company did listen. Not coincidentally, this was the same company that was further ahead than anyone else in developing autonomous vehicles for consumers.

That company was Google. And the reason they were listening

was because Larry Page and Sergey Brin deeply believed that this thing was possible. I'm not claiming that I did anything to change Larry's and Sergey's minds on the viability of autonomous technology. They were convinced of that long before I came along, as were Sebastian Thrun and Chris Urmson. Every member of the self-driving-car team was confident that autonomous technology would transform the way human beings managed their transportation.

However, as an adviser to Google, I helped key members of the Chauffeur team grasp the scale of the revolutionary effects of the approaching disruption. Bill Jordan and I presented the results of our research in Mountain View on December 12, 2011. About twenty-five members of Google's self-driving-car team attended the talk. My colleagues perked up when they saw the driver-time cost savings. Bear in mind, these were people for whom time was very valuable. Mostly they were engineers or programmers. Their productivity was important to their self-esteem. Not only that—they lived in *California*, in Silicon Valley, home to some of the world's most traffic-snarled and boring commutes. These men and women had all sat on nearby freeways fretting over the time they were wasting.

Another thing that interested them was how small the fleets could be to establish their utility. You needed to service only about 10 percent of the travelers in Ann Arbor to reach economies of scale, I said. That meant that a shared mobility fleet would be comparatively easy to test. You could do it in areas where the weather stayed relatively warm and dry.

But what really interested them was our calculations about the size of the business opportunity. Let's assume, I said, that in the future, shared, autonomous fleets handled just 10 percent of total miles driven—which would have seemed a conservative estimate to the engineers in the room, all of whom were very high on the technology's potential. That 10 percent share equated to 300 billion miles per year driven in America. Assuming further that a profit of just $0.10 per mile could be realized, shared autonomous mobility could lead to an annual profit of $30 billion. That represented some of the best years of the world's most profitable companies, such as ExxonMobil and Apple.

The Chauffeur team had long known that autonomous transportation represented a sizable business opportunity. I'd spent many months working to get them thinking beyond autonomous vehicles, to a future that also included the other trends relevant to a mobility disruption. Some of them had already been thinking about the convergence of these trends. Now, together, all of us were grasping the enormous size of the market that could arise. We were growing to understand that an even bigger business opportunity existed if you began preparing not just for an autonomous future, but also for one that included shared fleets, electric propulsion and tailor-designed vehicles.

Bill Jordan and I eventually would release our research by posting it on the Earth Institute website. The figures formed a large part of a paper I wrote for the magazine *Nature* in 2013. An Organisation for Economic Co-operation and Development report would use our research as the basis for concluding just how transformative shared, autonomous mobility could be. And I would present the calculations around the world at numerous conferences and symposiums.

But the most important audience I had was the one that was listening on that day in December 2011 in Mountain View. "Your insights into transportation as a service, and its potential, were instrumental in setting the direction of our project," Chris Urmson would say to me later. That's kind of Urmson to say, but I'm not claiming that I sold Chauffeur on the idea of self-driving taxis. I *am* confident, though, that the seminar Bill Jordan and I conducted at Google in December 2011 underscored for the team the stakes of what they were working toward. If the Chauffeur team could commercialize safe autonomous technology, then our calculations suggested they really did stand to change the world.

IV

THE TIPPING POINT

Chapter Ten

THE STAMPEDE

Driving is the distraction.

—ALAN TAUB

Chauffeur developed the highway driver's assist product through 2011 and into 2012. By the fall of 2012, the product was ready to be tested. The one drawback was that even with the technology activated, the human driver still needed to pay attention to the road at all times, regardless of whether the feature was engaged. The driver could glance out the window but could not tune out, because the technology didn't handle all types of situations.

Let's say the vehicle came up on a new construction project with an unusual pattern of traffic cones. Or an accident—maybe a tractor-trailer had jackknifed across all lanes. These new, impossible-to-predict events were apt to confuse the software, triggering an alarm that signaled that the human operator needed to resume control of the vehicle. At first, the self-driving-car team set the software so that the vehicle provided six seconds of notice between the time the alarm sounded and when the operator would have to do something.

With the six-second cushion in place, the self-driving-car team arranged to dogfood the technology in the fall of 2012. "Dogfooding," in Silicon Valley–speak, refers to the sort of trial that occurs when employees, as a proxy for the public, begin using the company's own products. (Apparently the term originates from old Alpo commercials in which the actor Lorne Greene dished out the pet food for his own golden retrievers.)

Google employees, that is, men and women who'd never received any particular driver training and were not affiliated with Chauffeur, began using vehicles equipped with the technology in the early part of 2013. A camera installed in the test vehicle monitored what was happening. Soon after the self-driving-car team began reviewing the video of the human operators, they noticed some troubling behavior: The drivers were tuning out—and far beyond what was safe. One guy pulled out his laptop and did some work. A woman applied her makeup. But what convinced the team to halt testing was the guy who fell asleep, for an astonishing twenty-seven minutes, as he cruised along at 60 mph on the freeway.

In one sense, the technology worked *too* well. It lulled its operators into a relaxed state, which was a problem if the alarm to reengage sounded. The highway-assist driving product didn't cause any accidents, but the lack of driver focus it promoted seemed certain to cause an accident.

Which presented the team with another problem: How to keep the human operator's attention? How to keep them awake? How to ensure a driver didn't get so entranced by a movie on her iPad that she wasn't able to respond quickly enough when her attention was required? "We wanted to be *safer* than the human," Nathaniel Fairfield recalls. But, "if the human is the fallback system, you're never going to be safer than a human."

As the team began to develop ways to prompt the human operator to pay attention to the road, the highway driver's assist product began to seem like a digression to the Chauffeur team. They'd joined Google to create a new kind of vehicle, one that was safer than human-operated versions, one that transformed the world in all sorts of ways. And here they were, feeling like they were heading into a developmental cul-de-sac, creating a technology that didn't transform transportation, but rather made automobile commuting more convenient. Highway driver's assist wasn't going to solve traffic problems. Rather, if people zoned out too much and caused an accident, it seemed certain to actually create new ones.

One day Dolgov and Urmson were out on a highway somewhere, doing some testing in a vehicle equipped with the highway driver's assist product, and they happened upon an absurd metaphor that, to them, summed up their situation. "Imagine if you had discovered radiation," Urmson says. "Did you want to make glove warmers with it? Or do something really transformational, like invent a nuclear power station?"

The highway driver's assist product seemed like the glove warmer option.

It wasn't going to substantially increase the safety of highway driving. It wasn't going to transform the experience of automobile transportation. And the problem was tougher than they'd thought, not because the technology was difficult to develop, but because of an entirely human tendency to get bored, to seek out stimulation. Then they discovered that Mercedes was planning to release in its 2014 vehicles a feature, Distronic Plus with Steering Assist, that did much the same thing as Chauffeur's highway driver's assist.

For all these reasons, the Chauffeur team opted, soon after the trial began, to scrap the highway driver's assist project, so that they could concentrate on solving the entire problem: creating a self-driving car that could handle every aspect of an automobile's operation, from the time the human entered the vehicle until after he or she left. The only problem? Once again, they faced the issue that had caused them so much struggle two years before: How to bring this autonomous technology to market?

In December 2012, Chris Urmson gathered the roughly seventy people then working on the Chauffeur project and pointed us all in a new direction. We met in a conference room designed to accommodate about two dozen people. Every chair was taken. Engineers and computer scientists stood around the table. Those who arrived late created a bottleneck at the door. This sort of all-hands meeting was unusual for the Chauffeur project. We'd all known for some time that we'd opted not to pursue the highway driver's assist product. Now the expectation was that Urmson would point us all toward something new.

After two years consulting for Chauffeur I'd become comfortable with my role on the team. A lot of what I did in those early years was strategic. Similar to my position with General Motors, I was assigned to tackle questions that analyzed the way the future might play out, and the way the project might best position itself to benefit. Urmson and his fellow engineers also used me as a resource, as someone who came up from within the auto industry, for insight into how to approach various suppliers and even the major manufacturers about possible deals.

Another factor playing into my relationship with Chauffeur was my age. I was sixty-one in 2012, which meant that I had at least a generation on the engineers who formed the core of the Chauffeur team, most of whom were in their thirties or slightly older. At first I'd worried they'd regard me as an anachronism. But for whatever reason, they frequently sought my counsel. They respected that I'd spent a good part of my career trying to enact the same changes they were trying to make, from within the industry—and they seemed to enjoy hearing about the bruises I'd incurred along the way.

I grew particularly close to Chris Urmson, perhaps because I saw some of myself in him—not just because we shared a similar worldview. Urmson was just an easy guy to be around. Whenever I was in Mountain View, the two of us tried to get together for a one-on-one dinner, and we had a running thing going over where we would eat—he preferred Indian or Asian cuisine, while I always tried to go for Italian. You can tell a lot about a person by the way they treat their server at a restaurant. Urmson was unfailingly polite and respectful no matter who was on the other side of the interaction—which is, I think, a credit to his parents and upbringing. At the Chauffeur offices, whenever anyone manifested a Silicon Valley affectation, Urmson was always there with his dry wit to puncture the pretention, to remind us all how the attitude might play in Saskatchewan, where his folks lived, which also happened to be how the attitude might play elsewhere in the U.S. Plus, he was an engineer looking to reform the auto industry, facing down the same skeptics and naysayers that I had.

I also hit it off with Dmitri Dolgov, who'd done his PhD at the University of Michigan, and who impressed me with a presentation he conducted for the computer science department in Ann Arbor on self-driving cars. Dolgov's enthusiasm for his work reminded everyone of just how cool the Chauffeur project was. It wasn't exactly the world-transforming potential of the project that thrilled Dolgov, although that was part of it. Rather, I think the real origin of his excitement was the capacity for software, for computer programming, to pull off the feats that were just then growing possible. Dolgov had grown up in the eighties, when Atari consoles struggled to provide pixelated computer graphics, and here he was, using artificial intelligence to help teach an automobile how to recognize pedestrians in all their various permutations. The glee that he brought to his job was almost childlike, which made it both endearing and infectious.

Another one of my favorite people on the project turned out to be Bryan Salesky, whom I first met back when he was Urmson's software lead on Carnegie Mellon's Tartan Racing team in the DARPA Urban Challenge. Salesky and I started working for Chauffeur at about the same time. He'd grown up in Pittsburgh as well as suburban Detroit, which gave us some shared reference points. We bonded because of our similar approach to engineering. We both came at new challenges by first focusing on the big picture and then addressing the smaller details. We also shared an appreciation for the International Organization for Standardization guidelines—known as ISO standards because *iso* means "equal" in Greek. The guidelines can apply to processes pretty much anywhere. (Urmson's father, Paul, believes that he was the first person in the world to ISO-certify a prison when he applied the guidelines to a correctional-facility furniture factory in 1991.) Specific to the development of autonomous cars, they represented a method to manage changes to code and vehicles so that

everyone stayed informed and tracked the way any individual altera-
tions might affect the other parts of the system. Sounds boring,
right? And on some level, it is. Following the ISO guidelines while
developing an autonomous car requires lots of meetings in which
engineers brainstorm the various ways a system might fail—and then
devise ways to fix things, or avoid the failure in the first place. Salesky
also displayed a curiosity about the way things were done in the auto
industry that was unusual for people in Silicon Valley.

I loved working on the Chauffeur project. As an enthusiastic technologist I'd sometimes felt frustrated in Detroit, where convention
and tradition could be more important than innovation. Technology was often viewed with suspicion in the auto industry; in Silicon
Valley, it was accepted that technology would improve the world for
the better. And I felt this with my every interaction at Chauffeur.
When I'd left General Motors I'd feared I was entering a period of
retirement. Now, just a few years later, I found myself working on
one of the most exciting projects I'd ever encountered, with a great
team of idealistic people around me.

Problems did exist, however. In May 2012, after a little more than
a year with the project, Salesky came to me and told me he was
leaving Chauffeur. He missed Pittsburgh, but the bigger reason was
the problematic relationship between Urmson and Levandowski.
Having cooled after our August 2011 summit, the conflict was once
again affecting the team's ability to get things done and Salesky was
frustrated. This was a blow to the quality of the team, which really
benefited from Bryan's grounded approach to engineering and product development. I was disappointed to see him leave. Urmson and
Levandowski were just such different people. Levandowski operated
at a speed I'd never seen another human match. When he applied
himself to solving a problem, his productivity was extraordinary. But
the disadvantage of this character trait was that he grew impatient
when things didn't progress at the tempo he thought possible. And
if getting things done quickly required cutting a corner here or there,
well, those issues could always be cleaned up later.

In contrast, Urmson was more meticulous. The engineer was focused on safety. Numerous times we spoke about the benefits that autonomous vehicles could bring to America's roads. The farmer's math was compelling: Those 1.3 million roadway fatalities a year, worldwide, worked out to more than three thousand deaths a day. The project Urmson was leading could translate into a lot of lives saved. And if the intent was to save lives, Urmson was hyperaware that rolling it out before it was ready could be ruinous.

Let's say you were in charge of programming speech-recognition software for a virtual assistant. Correctly recognizing ninety-nine out of a hundred spoken words represented great functionality. A virtual assistant, like Amazon Alexa or Google Home, that correctly interpreted ninety-nine out of a hundred words could be a really valuable thing.

But an autonomous car that correctly recognized ninety-nine out of a hundred stop signs would be awful. Not only could that one-out-of-a-hundred mistake result in someone's death—it also would result in the discrediting of a technology that many in Detroit already were criticizing as unsafe. If Levandowski's instinct was to just get the tech out there, to roll it out as fast as possible, Urmson's instinct went in exactly the opposite direction. He devoted Chauffeur's resources to whipping up a simulation team that would run the software in virtual reality, confronting the program with thousands of permutations of events, just to see what might trip things up and correcting things when a problem arose. Similarly, he set up a real-world version of these simulations at a 1,400-acre decommissioned military facility, the former Castle Air Force Base, about one hundred miles east of Mountain View. There, dozens of people, known as the Orange Team, devised scenarios that might bug up the self-driving software in the real world. For example, as one of the vehicles approached an intersection, an Orange Team member carrying a wall-size canvas might step onto the street. Would the software recognize the device as a pedestrian or some other object, like a truck? What about a cyclist who happened to be carrying an enormous net bag of beach balls—would the car be able to understand that, despite the unusual silhouette, the entity would behave like a cyclist?

Such steps required time, and money. They struck me as exactly the right moves. And they were illustrative of a rigorous approach on Urmson's part that drove Levandowski crazy.

It was, in many ways, an impossible situation. Consider that Levandowski reported to Urmson. But in another respect, his larger share of the Chauffeur Bonus Plan made him act like the company's biggest shareholder—one who had lost faith in the ability of the company's CEO to maximize the company's value, which Levandowski saw as directly affecting his financial hopes when the team's stake in the plan vested.

What evolved was a poisonous dynamic. Urmson would say one thing, then Levandowski would criticize it to the various members of the team to discredit him. In general the people on Chauffeur liked Urmson. They respected his leadership and enjoyed working with him as a person. But Levandowski was a brilliant manipulator who was adept at finding exactly the right weaknesses in a plan to mention to just the right engineer. The negative dynamic between the two men sapped morale.

"It was pretty bad," Larry Page would say in court documents, years later, referring to what he calls the "fractious relationship" between Urmson and Levandowski. "I think they had a really hard time getting along, and yet they worked together for a long time. And it was a constant—yeah, constant management headache to help them get through that . . . [Levandowski] clearly felt things could be done better." At the same time, according to Page, "There were a lot of concerns about Anthony . . . People were definitely concerned about trusting him." Certainly, one of those people was Chris Urmson, who would observe years later that Anthony was a "lost cause" due to his "manipulations and lack of enthusiasm and commitment to the project."

To anyone who knew the size of the stake he had in the plan, Levandowski's intransigence might have been puzzling. One take on the rational thing to do in his situation would be: Shut up, set aside your doubts and your ego and do what's best for the team. But that wasn't Levandowski, and this was, perhaps, a singular situation. This was a project that could improve the world in so many ways. And remember, my calculations suggested the market for shared, autonomous and electric mobility could be worth trillions in the United States alone. Levandowski may have felt that the size of this opportunity was so big, and his position as one of the handful of people in the world who understood LIDAR and its application to autonomous driving so singular, that he could make additional money by going out on his own and competing with Chauffeur. It was just a few years later that it would be alleged that Levandowski was a prime mover behind another company known as Odin Wave, apparently founded to develop LIDAR technology—which also happened to be one of his job responsibilities on the Chauffeur team.

— — —

Urmson's December 2012 announcement to the Chauffeur team of a new direction had a lot to do with events then happening in the San Francisco Bay Area. Earlier that year, in February, a tech entrepreneur named Sunil Paul, who had already enjoyed earlier success founding and then selling to Symantec an anti-spam company named Brightmail, began arranging rides in San Francisco via a mobile app called Sidecar. Paul had long been inspired to try to improve the existing transportation system. He'd filed a patent in 2002 for an algorithm to work out an efficient route from place to place. More recently, Paul was inspired by Uber's black car–hailing software—then spreading from San Francisco to such cities as Washington, D.C., and Chicago—and consequently, a hot topic in Silicon Valley start-up culture. But the major innovation with Paul's Uber-inspired ride-sharing app was that it allowed *anyone* to provide the ride, not just black cars or taxis. Anyone who had an open seat in a car could help out anyone else who needed a ride. (There was some paperwork. Ride providers had to pass a background check and have a valid license and insurance.) In following with Paul's idealistic vision to use technology to improve transportation for all,

Sidecar didn't even charge a mandatory fee at first. Rather, users were encouraged to provide the driver with a donation, from which the Sidecar app took a 20 percent commission.

Just three months after Sidecar's debut that May, Zimride leaders Logan Green and John Zimmer rolled out the beta testing for their own anyone-to-anyone ride-sharing app, which they referred to as Lyft. Every Lyft driver signaled their willingness to provide transportation by fixing a massive stuffed pink mustache on the front of the car. "To live in San Francisco in the year 2012 was to wonder, with mounting curiosity, why those weird pink carstaches were suddenly everywhere," writes Brad Stone in his book on Silicon Valley's disruptive new businesses, *The Upstarts*, noting later that "California was ground zero in the ridesharing movement."

Silicon Valley works in waves. Dating back to semiconductors and the computer chip companies founded in the sixties, to the late-nineties dot-com boom, and the millennial mania for social media, the venture capital and media attention seem to glom onto a single topic at a time. And for a moment in 2012, the hot topic was ride sharing, and anything else that used tech to attack the waste in the one-car, one-driver personal transportation model. Lyft, Sidecar and a French company named ticket2go were setting up business in the Bay Area. And while Lyft's explicit desire was to "replace car ownership," Uber watched with interest, eventually rolling out its own provided-by-anyone ride-sharing service, UberX, in January 2013.

Of course, the Chauffeur team paid attention to what was going on. Everyone recognized the ride-sharing model was a step toward the sort of on-demand mobility that we figured was the end goal of our work at Chauffeur. "We thought, hey, we're in a unique position," recalls Nathaniel Fairfield. "We know things people don't know. We've got a team that is second to none. We have the vision and the dream and the drive to go where we really want to go. Which is door to door—Well, let's fricking go there. Let's *do* that."

Urmson wanted to pursue the development of a limited-production-run prototype vehicle that would extend the innovations created by Sidecar and Lyft. He envisioned a world of driverless taxis zipping about cities, picking up passengers, providing rides, then setting off on the next call—a business that within Chauffeur was referred to as "transportation as a service." That fall, Urmson began speaking with Ron Medford, deputy administrator of the National Highway Traffic Safety Administration, the number two person in the federal government bureaucracy charged with ensuring the safety of the nation's automobile transportation system. Medford was a member of my generation. His wire-framed glasses, close-cropped hair and business-casual dress reflected a career spent working in Washington and Detroit, shuttling between the car manufacturers and government in an effort to reduce auto-crash-related deaths. Within NHTSA, Medford had pushed to raise awareness of the dangers of distracted driving. He also helped to institute more stringent fuel economy standards for American automakers. Urmson heard that Medford was about to retire from NHTSA and saw an opportunity. Someone with Medford's knowledge of state and federal driving regulations could be valuable to Chauffeur. Urmson hired Medford in November 2012 in a widely reported move that was interpreted by the tech media as another indication that Google intended to make autonomous cars a reality. I thought hiring Medford was a great move by Urmson. Medford's position on the Chauffeur team was director of safety. He was to work with state and federal governments to craft new regulations around autonomous vehicles.

Now the Chauffeur team lead asked Medford how the U.S. regulatory framework would deal with a fully autonomous car transporting passengers. Rolling out some sort of service with the kind of Toyota Prius sedans and Lexus SUVs that Chauffeur was using for testing would be difficult, Medford said—but had Urmson considered pursuing a low-speed vehicle?

That perked up Urmson's ears.

Motor vehicles are regulated by state governments, and most states included a separate legal category for low-speed vehicles, which fall somewhere between a golf cart and a conventional car. Regulations varied from state to state, but in general the law tended to define a golf cart as a four-wheeled vehicle with a maximum speed of 15 mph. A golf cart doesn't require a license plate, doesn't need safety equipment like headlights or turning indicators, and generally isn't allowed on public roads. A low-speed vehicle *is* allowed on public roads. Most states allow LSVs on public roads with posted speed limits up to 35 mph. They can reach speeds of 25 mph, have a weight limit of three thousand pounds, require a driver's license to operate and even have a vehicle identification number. Medford said it would be a lot easier to get a low-speed autonomous vehicle cleared to operate on public roads in California than it would be for the Toyota Priuses and Lexus SUVs that Google already owned.

In the conference room on that day in December 2012, Urmson told the team that he wanted to pursue on-demand mobility as a business model—a vision of the world where personal mobility in urban centers would be dominated by, essentially, a driverless Uber or Lyft model. The service could be billed monthly, weekly, annually or on a per-use basis. Most trips would happen in vehicles that weren't owned by the riders. Their shared use would be geared to the way people traveled.

Urmson wanted to create a vehicle designed specifically for use in transportation-as-a-service fleets. From the first DARPA challenge, the engineers who worked on the problem devoted themselves to adapting existing vehicles to autonomous purposes. The process entailed some compromises. The geometry of conventional vehicles created blind spots for the sensors required for autonomy. Now Google's self-driving-car team would design an automobile to be *explicitly* autonomous. At the meeting, Urmson set out some of the stipulations he wanted Google's first-ever clean-sheet vehicle prototype to pursue. He wanted it to be battery-electric. He wanted it to be friendly to passengers as well as the people outside the vehicle. That is, he wanted it to be considerate to vulnerable road users such as cyclists and pedestrians. He wanted the vehicle to be designed so that if it hit a pedestrian, the person could walk away.

I thought the new direction was tremendously exciting. The task was similar to the work that I'd pursued at GM, and the car that Urmson envisioned sounded as though it would have a lot of similarities to the GM EN-V mobility pod. Strategically, too, it was smart. Very few people on the team had auto-manufacturing experience. Designing and building this vehicle was a way for the team to learn about the process and to see if it was, in fact, as hard as Detroit made it out to be.

Several months after the meeting, Urmson called his old friend Bryan Salesky back to Silicon Valley. "We figured it out. We figured out the plan. I'm in charge," Urmson said, referring to the tension between Levandowski and Urmson. "The dynamics are fixed." Salesky became the lead on the development of the overall vehicle, which previously had been Levandowski's role. Dmitri Dolgov took the lead on the software development. And Levandowski was essentially relegated to leading the engineering of the autonomous hardware, such as the LIDAR and radar sensors.

What followed was a fascinating design exercise. How should a vehicle look if it was intended expressly to provide transportation as a service in a city center? Urmson worked with Chauffeur design lead YooJung Ahn, whom everyone called YJ. Born and educated in South Korea, YJ attended Hongik University in Seoul, then pursued her master's at the Illinois Institute of Technology, subsequently working for Motorola and LG Electronics. She regarded her lack of automotive experience as a strength on this job, because she wanted to pursue the problem from the perspective of the people who would ride in the vehicle—most of whom had very little industry experience themselves. At my encouragement, Urmson also brought in David Rand, former executive designer for General Motors, to help advise YJ. Rand had overseen the design process that had created the GM EN-V prototypes.

First-time users would initially be a little anxious about riding in an autonomous car. So what sort of a vehicle would calm that anxiety? YJ and Rand's design could be perceived as a response to the 2011 Dodge Charger commercial that had portrayed self-driving vehicles as the first step toward a *Terminator*-like future of robot overlords. They came up with a simple, clean and fun aesthetic. In contrast to Arnold Schwarzenegger's scary red-eyed robot, the curved lines of the vehicle exterior created an organic feel to the car.

Because Urmson hoped it would liberate transportation for those who couldn't drive—the elderly or disabled, for example—the vehicle needed to be easy to enter and exit. To achieve that, the floor would be flat, and not too far from the ground. When the conversation turned to the placement and feel of components such as steering and brakes, the team made a radical decision. The sort of steering wheel the Google mobility pod would have was *no steering wheel at all*. Why did it need one? Not having any controls felt audacious—the sort of disruptive and futuristic step that should come from Chauffeur.

The interior felt roomy for such a small car, in part because the electric propulsion made bulky engine components unnecessary. There was a button to begin the experience, another to stop the vehicle in an emergency and a screen that displayed the time of day and the number of minutes remaining until the passenger reached the destination. Like a lot of vehicle front ends, this one's resembled a face, but one that was wide-eyed as well as wide-smiling. The future of transportation, Chauffeur's design said, was fun, accessible and inviting.

Once the rough details were established it was time for some major decisions. Among them, who would make it? Chauffeur ended up contracting Roush Engineering to create one hundred vehicles. My first test ride happened on the top floor of a parking garage adjacent to the Chauffeur facility. I loved the pod car the first time I rode in it. To me, the two-person pod was the epitome of the mobility disruption. The vehicle became known within Chauffeur as Firefly, which seemed suitable—after all, it would buzz around urban centers the way the little flying insects zigged and zagged their way through summer nights.

— — —

I was excited about the development of Firefly. Yet I was also anxious. Sometimes I'd wake up in the middle of the night and I'd worry. I envisioned all sorts of scenarios. Google was beginning to work with state regulators to set the stage for legislation that would allow the safe operation of autonomous vehicles on public roads. We were hearing indications that the auto industry's advocacy organization was lobbying hard against us, to slow things down. Would Detroit's hostility to our efforts somehow affect things inside Google and lead to the elimination of Chauffeur's funding? I'd spent much of my career managing promising technological projects. I'd seen research efforts that seemed destined to lead to major economic and environmental benefits go nowhere because of the change-averse mind-set of the automotive industry. Hydrogen fuel cells, for example. I shouted myself hoarse while at GM, talking about the benefits their widespread automotive use could provide to the world—and still, nearly two decades after I'd first begun thinking they would eliminate oil dependence and the emission problem from automobiles, they remained a niche technology. Could autonomous vehicles, and the potential of a widespread mobility disruption, suffer a similar fate?

I coped with my fears by setting out to tell as many people as I could about the effects of the technology. My consulting work with Google was supposed to take about 20 percent of my time. I frequently worked with other clients in the energy, insurance, real estate and logistics industries on projects related to the future of transportation and how to respond to it. And then in between my consulting work I would give presentations about my vision for the future of mobility at dozens of conferences all over the world, from Paris to Hong Kong, Australia to Canada. I racked up personal high scores in frequent-flier miles as I described all the changes I thought widespread adoption of transportation as a service might bring.

These presentations were designed to provoke the auto industry into noticing me. Sometimes I began my slide presentation with that famous "see no evil, hear no evil, speak no evil" image of the three monkeys, which was my take on the stance of the auto industry. The safety benefits of self-driving vehicles stood to eliminate at least 90 percent of the 1.3 million roadway fatalities that happened around the world, I'd say. Delaying things by a single day basically entailed killing somewhere on earth an additional 3,000 people. "This is a transformational opportunity," I'd say, describing the Earth Institute calculations about the potential cost savings for consumers and the size of the market for the technology. I'd also describe the new business models I saw dominating a world of disrupted mobility, in which the automotive industry would move from selling vehicles, gas and insurance to selling miles, trips and experiences. And I drew an analogy to President John F. Kennedy's 1961 pledge to send a man to the moon. That was an example of audacious thinking that changed the world—and now, I'd say, we had to think audaciously about the advent of transportation as a service.

Not all the changes represented improvements, I realized as I considered how to encourage the auto industry to disrupt itself. Driving through my own home state of Michigan, I would get a sinking feeling as I considered the effects of society embracing on-demand mobility without widespread governmental programs designed to ease the effects of job loss. The adoption of these driverless taxis in cities around America, and the world, does seem likely to trigger significant labor effects. Let's start with the implications for those employed as drivers. According to Bureau of Labor Statistics survey data, about 4 million people in the United States work as some sort of driver. That includes 1.7 million heavy-truck or tractor-trailer operators, 685,000 bus drivers and 189,000 taxi or limo drivers. Such drivers represent nearly 3 percent of the American workforce. What will all those people do once vehicles operate themselves? Once Uber uses driverless taxis, once Domino's delivers pizzas by robot and UPS parcels move from the logistics center to the home without help from humans?

Politicians, unions and the individuals who stand to lose their jobs may be tempted to resist the disruption precisely because they fear its effects. I get that. If my profession were one that was vulnerable, there's no amount of rationalization that would make me comfortable with it. But consider what stopping things would mean.

Let's say the 4 million people employed as drivers work 40 hours a week for 50 weeks a year. That translates into 8 billion hours of paid work that we stand to lose, thanks to the widespread adoption of autonomous on-demand mobility.

Now recall that America's 212 million licensed drivers operate their vehicles an average of 56 minutes a day. As a nation, that translates into approximately 72 billion hours that people in our nation spend driving every year. Are we really going to stop something that stands to liberate 72 billion hours a year, which many of us could use to work additional hours at our jobs, because we stand to lose 8 billion hours of paid driving work?

There's another way that American labor could be affected by the mobility disruption. Recall that the driverless taxis will be on the road much more than today's personally owned gas-powered vehicles. Consequently, the shared driverless vehicles will accumulate miles much faster. Let's say today's vehicles wear out after 10 to 15 years and the accumulation of 150,000 miles. Some of those parts wear out from *age* rather than *use*. That is, they wear out because they're 10 or 15 years old, not because they've been driven 150,000 miles. The shared, autonomous vehicles will burst through the 150,000-mile barrier in just a couple of years—and they'll keep on driving. Plus, electric vehicles are much simpler to design and engineer, which means they'll have far fewer parts to wear out, as Byron McCormick made clear to Rick Wagoner and me at GM. Taking into account the electric powertrain and the effects of higher use rates, my data suggest the post-disruption vehicles will last 300,000 miles accumulated in just four or five years. Which, in turn, affects how many vehicles are manufactured. My own calculations suggest we could get by with about half the vehicles we use today—and the ones we do have will be lighter, with fewer parts. All that translates into a much smaller automobile manufacturing sector.

Finally, the sector of our economy that provides services to car owners may contract. When an enormous proportion of automobiles are maintained by fleet providers rather than individual Americans, we won't need as many gas stations or car washes. Or mechanics shops, brake-repair specialists or while-you-wait oil-change depots.

Easing my anxiety about the changes ahead is the knowledge that we as a nation already see a lot of flux in the labor market, with turnover—defined as the number of times Americans quit, are laid off from, or are "discharged" from their jobs—amounting to 5.2 million jobs per month, according to the Bureau of Labor Statistics. More than the mobility disruption will affect work in the years ahead. Automation will impact almost every economic sector. Industrial robots perform 10 percent of the tasks in manufacturing today, according to a study by the Boston Consulting Group, which predicts the proportion of tasks handled by robots will reach 25 percent by 2025. McKinsey & Company, the consultancy, conducted a survey on American jobs to learn how many will be eliminated by robots, machines or artificial intelligence. "Overall," says McKinsey, "we estimate that 49 percent of the activities that people are paid to do in the global economy have the potential to be automated by adapting currently demonstrated technology."

Years ago, you got into an elevator and a full-time technician, whose entire job was to ferry passengers up and down to various floors in the building, operated the machinery to get you to the level you wanted. Banks featured many more tellers, whose jobs are now done by account holders themselves via ATMs and Internet banking. More recently, parking lot attendants have been replaced by automated payment machines. In 1800, the proportion of the total American labor force that worked in agriculture was near 80 percent. In 2000, thanks to the effects of machines such as harvesters and combines, agricultural labor formed less than 5 percent of the workforce.

All these workers had to find new work, some of it higher paying and more pleasant than the way they used to make their livings. Similarly, new work will arise from the mobility disruption. One of the most forceful figures arguing this line is former Uber CEO Travis Kalanick. I disagree with Travis on many things, but on this one, some of what he's saying makes sense. "You know there was once a time you made a phone call and there was a person that, the operator, had to do switching, right?" Kalanick said in an interview with *Business Insider*'s Biz Carson, going on to mention the numerous workers who built telephone booths. "And then cellphones came and it is a beautiful thing, but then that created a whole new industry and all new kinds of jobs."

Some studies on contemporary labor disruptions actually suggest that incorporating technology into the workforce creates more work than it makes redundant. That McKinsey paper I cited previously references a 2011 study out of its French office that looked at the employment effects of the Internet. The report concluded that for every job the new technology eliminated, 2.4 new positions had been created.

What sort of new jobs might arise for the people who once worked as drivers? Lyft cofounder John Zimmer intends to evolve his on-demand mobility service to the point where the company will provide rides in what he describes as "rooms on wheels"—chambers where a concierge could provide a meal, or drinks from a bar, or assist with other services, such as a massage. Logistic technicians will oversee the computer-controlled deployment of vehicles to ensure the fleets are properly dispersed. Cleanliness will be an important differentiating factor in the market for shared mobility. The robots providing on-demand mobility will have to be vacuumed once a day, at least, requiring many more people to staff what used to be known as car washes.

Transportation as a service will spawn new work in other sectors, too. What are riders going to do for that 72 billion hours a year that suddenly gets freed up? Those who spend that extra time working will increase our productivity as a nation, possibly leading to further employment. Others will pass the extra hours consuming content. Better Internet infrastructure will have to be built and maintained to ensure that high-def video content can be delivered wirelessly to vehicles. Online retail will only become more prevalent after the mobility disruption. E-commerce was forecast by Forrester Research to have grown 13.3 percent in 2017, five times faster than that of bricks-and-mortar retailers. A remarkable half of all growth in e-commerce comes from Amazon—which as I write this is worth around $600 billion, placing it firmly in the world's top-ten biggest companies by market capitalization. The company and its e-commerce competitors seem poised for further growth in an economy that has fully adjusted to autonomous technology, because delivery costs will fall. It's expensive to pay a human being to drive goods to my door. And once you don't need the driver, you also dispense with the equipment the truck needs to keep the driver safe and comfortable: seats, a windshield, safety belts, the heating and air-conditioning, the metal and other materials required to build the passenger compartment.

Thanks to such changes, we can reduce the cost of over-the-road trucking by 50 percent. The area served within a day by large trucks able to drive twenty-two hours at a time, rather than the eleven-hour driving limit for human operators of tractor-trailers, doubles the single-day range of a single truck to 1,100 miles and quadruples the area that truck can service. Meanwhile, lighter and cheaper land- and air-based drones could deliver small packages, and the same driverless two-person pods moving *people* around could also deliver goods to homes and businesses at very low cost.

Yes, many people's jobs will be disrupted. And the risk exists that transportation as a service may contribute to the ongoing trend toward income inequality in the United States, allowing the wealthy to become even wealthier and the poor to grow ever more disadvantaged as income flows from small-business owners of gas stations and mechanics shops to mobility-on-demand fleet operators. A recent Brookings Institution policy brief raised the specter of a "Luddite backlash" triggered by "the accelerating level of labor market dislocation caused by widespread automation."

A Luddite backlash may already be happening, thanks to privacy concerns and misuse of social media data. But let's hope that the mobility disruption emerges unscathed. Rather than, say, legislation to stop transportation as a service, government should respond by passing public policy designed to ensure that the people the mobility disruption puts out of work fall *up*—into better jobs. Retraining programs would be one smart policy measure, along with the Brookings policy brief suggestion to provide displaced drivers a living wage for a transition period of up to twelve months. I'd also encourage tax policy to ensure the fluid movement of workers from areas of high unemployment to areas where the labor market is tighter.

In May 2014, Urmson and Salesky had several working prototypes of the Firefly concept vehicle and were preparing to tell the world about what they'd created. It had taken the team seventeen months. This was longer than the team's leadership might have hoped. That the development of the Firefly vehicle took so long, at least in the eyes of the Google leadership, helped to begin a transformation of the Chauffeur team's attitudes toward Detroit and the dominant American automakers.

Recall that Detroit was criticizing Google's self-driving-car development efforts. It's important to note that the hostility ran both ways. Early in the history of Chauffeur, both the team members and Google's leadership displayed a lack of respect for Detroit's automobile development efforts. Auto companies typically took around three years to develop a new car. At first, this astonished the guys on the Chauffeur team. *Three years?* they exclaimed, incredulous. *What on earth took so long?* Levandowski in particular epitomized this attitude, but it wasn't just him. In general, they attributed the long development cycle to Detroit's lack of competitive edge. The auto companies were lazy, Silicon Valley thought. Conservative, averse to new ideas and out of touch. They didn't know how to do innovation—at least, not the kind that might spur a societal disruption, the kind that had arisen from Silicon Valley numerous times in the last half century, whether we're talking smartphones or the Internet, personal computers or the transistor. Many of Chauffer's team, for example, believed that Henry Ford was a remarkable innovator but that somewhere along the line, that spirit had withered in Detroit. They believed that most of the American engineers who continued Henry Ford's spirit worked in Silicon Valley.

The Firefly development effort taught Chris Urmson and Bryan Salesky a great deal. They gained an appreciation for what took the traditional manufacturers so long. Designing a vehicle was comparatively easy. What was *difficult*—which was also the part of the process at which Detroit engineering talent excelled—was a process the automotive industry referred to as "hardening" the vehicle's various components. Hardening meant engineering every part so that it worked in every condition that might confront the automobile for longer than a decade, as it was driven along 150,000 miles: in Seattle rainstorms, Arizona desert, Minnesota cold snaps, North Carolina gales and even Gulf Coast hurricanes. Sure, Chauffeur developed Firefly in about half the time it takes Detroit to develop a new-production car. But consider that they planned to make only about one hundred Firefly vehicles, all of them the same, rather than hundreds of thousands of vehicles ranging from the standard, no-frills edition to variants loaded with fancy options. Having gone through the development of Firefly, Salesky and Urmson began to regard Detroit with newfound respect.

As the first few months of 2014 passed, I tried, and failed, to convince Urmson to temper the messaging Chauffeur was planning to roll out around Firefly. I did not think it wise for the team to go public with the fact they'd stopped pursuing the highway driver's assist product. Why? The rest of the automotive industry was still working on various versions of highway driver's assist projects. Many of the efforts relied on products from a single Israeli company, Mobileye, cofounded by a former MIT computer-vision expert, Amnon Shashua. The products did not rely on LIDAR because Shashua believed the scanning devices were too expensive. "You cannot have a car with $70,000 of equipment," Shashua told the *New York Times'* John Markoff, "and imagine that it will go into mass production." Consequently, for companies like Volvo and Nissan, Mobileye created systems that enabled something called "traffic jam assist," which kept the car driving itself safely in the lane of a highway so long as the driver kept contact with the steering wheel. The company also provided obstacle-detection capability, which could slow down the car to avoid pedestrians and cyclists. Tesla's Elon Musk initially made overtures to Google to partner on self-driving technology but ultimately made a deal with Mobileye.

Chauffeur did not think much of the Mobileye approach. Nor did I. Since it didn't use LIDAR, the Mobileye product essentially relied on a computer chip comparing camera images to an enormous data set of pre-loaded pictures to recognize the sorts of obstacles the car might encounter. It worked fine, usually, on a limited-access highway, so long as the human was prepared to take over. But to create a vehicle that could navigate itself safely door to door, through traffic lights and roundabouts and all-way stops, the Chauffeur team believed the vehicle had to feature LIDAR, which in turn made it much easier to use high-resolution digital maps as reference points.

I thought we should let everyone else pursue their highway-driver's-assist wild-goose chase. Telling the world that Chauffeur was pursuing door-to-door autonomy—was so serious about it, in fact, that the team was developing prototypes that had neither steering wheels, brakes nor accelerators—would just get everyone else thinking about pursuing the same thing. Plus, it was easy to connect the dots from Firefly to the driverless-taxi, transportation-as-a-service business model. I mean, what else could the vehicle be used for, besides ferrying passengers around urban centers? I thought it would be much better if we waited and disclosed news of the Firefly project only when a transportation-as-a-service product was up and ready. Why, after all, would we ever alert potential competitors to our strategy before we were ready to deploy it?

But Google's tendency was to be more transparent than my Detroit-bred instincts. To make the vehicles safe for humans to ride in, they had to be tested on public roads. People would be discussing them regardless. Besides, the self-driving-car team was so far ahead of any and all competitors that they couldn't conceive of a future where anyone would ever catch up to them. Why *not* tell the world about the project? Perhaps the publicity would help convince automakers that it made sense to work with Chauffeur.

What no one anticipated is that the disclosure of Firefly's existence would begin a sequence of events that would result in a stampede by companies from numerous different corporate sectors into the market for transportation as a service. The news of the pod car was set to be revealed for the first time at the end of May 2014 by Sergey Brin at the inaugural Code Conference, a tech-industry gathering staged at a resort on the shores of the Pacific Ocean in the Los Angeles suburb of Rancho Palos Verdes, organized by legendary tech journalists Kara Swisher and Walt Mossberg. To generate buzz around self-driving cars in advance of the conference, Google staged a press event in Mountain View on May 13 featuring rides of about a mile long, from the Computer History Museum to Google's corporate headquarters, in self-driving Lexus SUVs. Next, Google invited a handful of journalists to actually ride in the Firefly vehicle—the first time that any member of the public would see it. The team provided the ride to the tech journalists on the condition that they embargo their articles until after Brin disclosed the news at Code Conference.

On May 23, the demo began with the *New York Times*' John Markoff climbing into a self-driving Lexus with Chris Urmson (this would be Markoff's second time in a Chauffeur vehicle). The car drove them a few blocks to an isolated Mountain View parking lot—one that had been surrounded by Google security staff to ensure no pedestrians got a look at the Firefly vehicle.

Once they stood on the asphalt, Urmson took out his phone and activated an app that summoned the Firefly vehicle, which promptly rolled up with its LIDAR spinning busily atop it. Urmson and Markoff climbed in and buckled their safety straps. Then Urmson showed Markoff the button on the center console that would begin the ride. Markoff pressed it, and the vehicle successfully executed without incident several laps of the parking lot.

"It was a cross between riding in my office elevator . . . and memories of riding in the Disneyland Tomorrowland people mover as a child," Markoff would write.

After the demonstration, Markoff went back to Google to interview Sergey Brin. Markoff asked Brin how the company envisioned operating the Firefly vehicle, and Brin spoke of severing the relationship between "transportation and vehicle ownership."

"Regardless of Google," Brin told Markoff, "I think the right model for most of the world will be not through vehicle ownership . . . These should be provided as services for the most part."

Urmson also provided a similar ride to Swisher, who had just founded the tech-media website Recode, and her colleague Liz Gannes. Swisher was an interesting choice because she's one of the few reporters covering Silicon Valley who speaks truth to power, in her writing and in person, regardless of how many billions are in the bank account of her conversation partner. Then married to Google executive Megan Smith, with whom she had two children, Swisher was famous for despising the search giant's Glass product and was reputed to be one of the toughest interviewers in the valley. She was certain to be honest in her reaction to Firefly.

Swisher seemed delighted with the look of the vehicle—an ovoid, organic design with anthropomorphic headlights and a black sensor panel at the front that resembled a koala's black nose. Moments later, Swisher and Gannes were zooming off for their laps around the parking lot. They seemed thoroughly impressed by the experience. "It's delightful," Swisher said. "This is pretty cool, I have to say . . . Very nice."

On May 27, at the Code Conference, Brin took the stage in a T-shirt, loose pants and a pair of Crocs. On his face was the Google Glass device so loathed by Swisher. The first few minutes of the interview amounted to small talk. Then Brin disclosed the news of the self-driving car by showing video of the demo conducted by Swisher.

"What you're introducing, this thing, does it have a name?" Swisher asked.

"This is just our prototype; it doesn't have a name because it's still in the prototype stage," Brin said, keeping the Firefly code name secret. "The key thing is it doesn't have a steering wheel . . . We took a look from the ground up. If we had self-driving cars in the world—what should they be like?" Brin said.

"How does the world look with these cars?" Mossberg asked.

"The reason that I'm excited about these self-driving cars is their ability to change the world, and the community around you," said Brin, who went on to describe the way the vehicle stood to change the mobility of people currently underserved by conventional transportation infrastructure, such as impoverished rural residents, the young, the old and the disabled.

"The main reason that I wanted to—that we collectively, the team and I, decided to develop this prototype vehicle, is we can do a better job than adding on to an existing vehicle," Brin said. "The experience feels different. You're just sitting there in the front. There's no steering wheel, no pedals. For me it was actually very relaxing . . . Ten seconds after getting in I was doing email. It ultimately reminded me of catching a chairlift by yourself. There's a bit of solitude that I felt very enjoyable."

One of Brin's main talking points concerned safety. Designing the car to be autonomous allowed better sensor placement, which made the vehicle safer because it could better sense its environment. He mentioned the way the steering and the brakes were controlled by two redundant systems, and the safety benefits of limiting the vehicle to just 25 mph—particularly since this all allowed for two feet of protective foam on the car's front end.

Brin was far less descriptive in his answers to anything that involved what Google planned to do with the vehicle. "Does Google want to be a car company?" Swisher asked.

"We worked with partners to build this prototype. We expect we'll work with partners in the future," Brin said, mentioning as one possible template the way Google outsourced the actual manufacturing of the Nexus phone. "The reason we want to talk about today is you'll see these driving around," Brin said. "It's still early . . . Being broadly available is still a ways away."

In retrospect, what was fascinating about Brin's interview with Swisher and Mossberg was what he *didn't* say. Brin spoke only in general terms about Firefly—saying just enough to disclose the project's existence, but sharing very little about Google's long-term plans. Urmson was more candid in a May 29 conference call with media, according to a transcript posted to the web by *The Atlantic's* Alexis Madrigal. "There are a lot of ways we can imagine this going," he said. "One is the direction of the shared vehicle . . . you can call up the vehicle and tell it where to go and then have it take you there. If you look at a vehicle purchase, it's the second largest purchase that most people make and it's a resource that sits idle 95 percent of the time, so it is a poor capital investment in some sense. So you can imagine if you could call one of these vehicles, have it take you where you want to go, and then have it go off to somebody else, the cost of transportation might be dramatically less . . . we think it's pretty exciting."

The key question in Brin's interview came from Swisher. "Where does Uber fit in all this?" she asked. The question was relevant because Google had the year before invested $238 million in the ride-sharing firm through its investment arm, Google Ventures, and placed the search giant's chief legal officer, David Drummond, on the Uber board of directors. Meanwhile, Firefly's so-called pod car seemed to set the company up to be entering some sort of a ride-providing service. Swisher was asking whether Uber would be involved—or was Google positioning itself to *compete* with the San Francisco ride-sharing giant?

Uber CEO Travis Kalanick was trying to figure out the same thing. Both Kalanick and Drummond also were at the Code Conference, with Kalanick due to be interviewed by Swisher the day after Brin. Soon after the Brin interview, Kalanick confronted

Drummond in a face-to-face encounter. Kalanick wanted to know whether Google was positioning itself to cooperate or compete with Uber. "I recall him just expressing concern," Drummond said in a deposition. Drummond told Kalanick that some provision of transportation as a service was one of the things Google was considering. "In any event, any competition was likely quite a ways off and that we—but we—that we should talk about these things as things developed," Drummond said.

Markoff's article about the new Google car was posted to the *New York Times*' website late in the evening on the same day as Brin's Firefly announcement—and in it, Markoff conducted some adroit speculation about Google's plans for the vehicle. The veteran tech reporter led with the news that Google was designing a car that didn't have a steering wheel or brake or gas pedals. "The company has begun building a fleet of 100 experimental electric-powered vehicles that will dispense with all the standard controls found in modern automobiles," Markoff wrote.

"Although both Sergey Brin, a Google co-founder involved with the project, and Dr. Urmson were coy about what the search engine company wants to do with its self-driving cars, I think the answer is now clear," Markoff wrote in a related post on the *Times*' tech blog, Bits. "One potential use: driverless taxi cabs." Describing the implications of that, Markoff mentioned my Earth Institute paper, summing up the way Bill Jordan and I had calculated that a "futuristic robot fleet" could provide mobility in Manhattan comparative to the borough's taxicabs, with a shorter wait time and at just a fraction of the cost.

One of the people monitoring this coverage most intently was Kalanick. Anyone in his position would have been wondering about Google's intentions. According to court documents, the human driver accounts for approximately 70 to 90 percent of Uber's cost per mile. The Uber CEO regarded the prospect of on-demand mobility as similar to search, or social media, in that it was a technology that benefited from network effects—the more people who used the service, the more compelling the service would be. If that was true, the initial leader would have a big advantage over others, who entered the market later. In Kalanick's mind, driverless taxis was a trillion-dollar prize in a winner-take-all event. His testimony in the Uber versus Waymo trial depict him and, by extension, his company as a little brother eager to hang out with an adored older sibling. And by "hang out" I mean cooperate to become the world's largest provider of on-demand mobility. By the time he was interviewed by Swisher, Kalanick had already ridden in one of the search giant's self-driving cars. The year before, one had picked him up to whisk him to a meeting with Larry Page. At the meeting, Kalanick tried to discern what Google planned to do with its technology, and actively pitched cooperation between the two companies. But Google's executives did the same thing each time Kalanick brought up the matter: They rebuffed him, saying it was too early to make any commitments. Now it was looking to Kalanick like one of his major investors was positioning itself to compete with his company.

Kalanick was characteristically pugnacious in his Code Conference interview with Swisher. At first, the subject of his ire was the taxi industry. "When we started Uber it wasn't about a battle or a war," Kalanick told Swisher. "But that war or battle has been brought to us. I think for too long we were tech geeks that didn't realize the battle was happening . . . We didn't realize it but we're in this political campaign, and the candidate is Uber. And the opponent is an asshole named Taxi . . . Nobody likes him, he's not a very nice character, people don't like what he does, but he's so woven into the political fabric and machinery that lots of people owe him favors and he keeps paying so the political machinery likes him . . ."

"What did you think of the self-driving car?" Swisher asked Kalanick.

"Love it," Kalanick said. "All day long. I mean, look, I'm not going to be manufacturing cars. Like, that's not what I plan on doing. Somebody's got to make them. And when those bad boys are made— Look, the magic of self-driving vehicles, is that the reason Uber could be expensive, you're not just paying for the car, you're paying for the other dude in the car."

"Who's driving," Swisher clarified.

"Who's driving, that's right," Kalanick said. "And so, when there's no other dude in the car, the cost of taking an Uber *anywhere* becomes cheaper than owning a vehicle. Even if you want to go on a road trip, it would just be cheaper. And so the magic there is that you basically bring the cost down below the cost of ownership. For *everybody*. And then car ownership goes away."

"People are such a pain in the neck," Swisher observed irreverently.

"And of course that means safer rides, more environmentally friendly, it means a lot of things."

"Google has a big stake in you," Swisher said.

"They have a *small* stake in us," Kalanick clarified, then just weeks away from announcing a new round of investment that would value the company at $17 billion, making it the world's most valuable privately owned company.

"Would you sell to Google to become their reservation system for this?"

"I get this question a lot."

"Well, you should," Swisher said. "I saw it for the first time, I turned to the guy and said, 'You need to buy Uber tomorrow.'"

"The way I think about it is," Kalanick said, "you just asked a happily married man who his next wife is going to be."

Swisher came right back at him. "So who's your next wife going to be?"

"I'm happily married, and that's very insulting, Kara," Kalanick said, deadpan.

That cracked up Swisher.

Later, in the Q&A that followed the interviews at the Code Conference, someone in the audience asked about the drivers. "You have a lot of dudes, and women, driving for Uber right now—they're probably not going to be thrilled that you're planning on replacing them with robots as soon as you can. How are you going to manage that transition from people-driven cars to driverless cars?"

"Look, the thing to keep in mind is, it's quite a ways off," Kalanick responded. "First—this is not something that is happening anytime in the near future. But if I were talking to one of the drivers that we partner with, I would say, look, this is the way the world is going to go. If Uber doesn't go there it's not going to exist." Which in 2014 was a remarkably prescient statement. Bear in mind, Google's Firefly news was just the second major public announcement in the company's effort to develop an autonomous car—the first announcement being the existence of the development effort back in 2010. And the day after Brin made the disclosure, Kalanick was saying that autonomous cars were *inevitable*. Not only that: The dominant business model would be the driverless-taxi structure that Chauffeur was gunning toward. And because that model would provide comparable mobility freedom without the cost of a human driver, it could price out of existence any competitor that kept its drivers. As Kalanick said, *If Uber doesn't go there, it's not going to exist.* The Uber cofounder would later speak of autonomous technology as an existential threat to his company. This was the first version of that observation.

"The world isn't always great," Kalanick said, finishing his thought. "It's the way of the world. It's the way of technology and progress. Unfortunately we all have to find ways to change with the world."

Then Kalanick turned the issue around. "Imagine if I'm talking to a rider," he said. "And I told them that they're going to be able to get anywhere they ever want to go at *a tiny fraction* of what it costs to own a car. And what if I told them that congestion on freeways is going to go away." He mentioned the way new cars will be safer, too. "That's kind of a big deal. And so—it's not technology for technology's sake. There's real advancement for how cities work, and how people *move* in cities, that is going to change our lives."

Kalanick's Code Conference interview caused some hubbub on social media. It was startling to hear a CEO speak so bluntly. Most of the comments involved the way he characterized Uber's work as a battle against "an asshole named Taxi." A good part of the response came from drivers, and people who sympathized with drivers, whom Kalanick had essentially declared redundant. The response prompted him to clarify in the conference's wake that he thought driverless cars were at least ten years out.

In the months after the conference, Kalanick seems to have weighed how exactly to respond to Google's announcement. He hoped the best solution was for the two companies to cooperate; to gun hard for the transportation-as-a-service model, together. A single entity that combined Uber's deployment expertise with Chauffeur's autonomous technology would have been a potent combination. But the most his entreaties to Google got him was a second ride in Chauffeur's self-driving car in September 2014. Urmson himself rode along with Kalanick, the first time the engineer had met the Uber CEO. "He seemed really, obviously interested in it," Urmson recalls. The technology had matured a lot in the four years since the public first learned of its existence. In good weather, on the well-mapped streets of Mountain View, the self-driving vehicle clearly was a better and safer driver than any human out there. Kalanick asked Urmson how long before the vehicles would be on the roads, providing rides to regular people. Urmson didn't give a precise figure but said that it would probably happen more quickly than most people thought. "I probably shouldn't have said that," Urmson would conclude later.

It's important to note that it wasn't the existence of the Firefly project that so alarmed Kalanick. Rather, it was the prospect that Google would deploy something like Firefly *itself*, operating its own ride-sharing service. The moment that Kalanick realized how urgent it was for Uber to develop its own strategy happened after the company's October 2014 board meeting, when Drummond pulled Kalanick aside for a private conversation. Drummond came right out and said it: Google was intending to compete with Uber in the ride-sharing space. Recounting the encounter later in a deposition, Kalanick said he responded with disappointment. "My face looked sullen," he said. "I don't remember my exact words, but I remember feeling disappointed a little bit, a little burned by the relationship" with Google. From Kalanick's point of view, the older brother had betrayed its younger sibling. Drummond stopped attending Uber's board meetings after that, thanks to the conflict-of-interest issue. However, Drummond and Kalanick continued to work to try to create a partnership between the two tech giants. "We both were trying to find a way for Uber and Google to partner autonomy and ride sharing together," Kalanick said. "[Drummond] was a big believer it was the right thing. I was a big believer it was the right thing . . . But David couldn't get . . . Larry and other folks and just Google, essentially, behind that notion."

Meanwhile, Kalanick monitored Chauffeur's every move. Urmson conducted a Q&A session with reporters while he was in Detroit for the North American International Auto Show the following January. Kalanick obtained a transcript of the session, and as soon as he went over it he emailed Drummond. "Urmson is openly discussing rolling out an autonomous vehicle ridesharing service," wrote the Uber CEO. "I'm thinking it's time to have a chat with Larry [Page] directly." Drummond agreed, but nothing happened until Kalanick emailed Drummond again on March 6. "Heard from a reliable source that Google will be starting a self-driving service in [Mountain View] in 3 months," Kalanick wrote erroneously. The meeting finally happened on March 10, over lunch at 1900 Charleston Road on the Google campus. In addition to Kalanick, Drummond and Page, Uber senior vice president of business, Emil Michael, also attended. In a description of the meeting, Drummond said Kalanick voiced his concerns about Google getting involved in ride-sharing. Page was up front and clear to Kalanick. Transportation as a service was one of the options Google was considering "in order to monetize, if you will, the self-driving-car opportunity," according to Drummond.

— — —

Kalanick wasn't alone in worrying about the effects of self-driving technology on his business. Quietly, steadily, people were beginning to listen—to grasp that this thing was coming, and when it did, it was going to transform not just the auto industry but society as a whole. Take John Casesa, one of the more likable guys I've come across in the milieu of financiers and engineers that dominate Detroit. At the start of 2015, Casesa was the senior managing director of Guggenheim Partners. He had begun his career at GM working as a product planner, then joined an investment bank as an analyst, writing well-reasoned assessments of the biggest and most important auto companies. Eventually he moved on to become a deal maker, specializing in mergers and acquisitions and other transactions for the automakers and the companies that supplied them, such as Toyota Motor Corporation, Magna International and Lear Corporation.

I thought Casesa was one of the top two or three smartest investment bankers working in the business. But the reason I liked him had little to do with acumen. He had a sparkle. Casesa always seemed to have this permanent grin about him. No matter who you were, he was interested in you and what you had to say.

At that point, Casesa had spent nearly thirty years in the auto industry, a business where automakers' success was defined by the extent they were able to sell lots of cars to lots of customers. Then Casesa read my Earth Institute paper—and realized that his business was about to undergo an important transformation.

Years ago, like many people in the auto industry, Casesa didn't really take the notion of self-driving cars or a large-scale mobility disruption too seriously. "People talked about it," Casesa recalls. "But I couldn't really make sense of it. It just seemed so . . . fantastical. I just couldn't think of it in terms of what it would do to the world—or business."

Then one afternoon, around the time Brin disclosed news of the Firefly vehicle, Casesa was up in the Glass House, the world headquarters of Ford Motor Company, located in Dearborn, Michigan, where the offices of the high-ranking executives sit atop the uppermost floors. Waiting to go into a meeting, with a few empty minutes to kill, he stared out the window at the view, which, appropriate to the setting, was dominated by a highway cloverleaf at the intersection of Michigan Avenue and Southfield Freeway. Cars were zooming onto the freeway and off the freeway, and as he watched, the investment banker grew pensive. How different this intersection would have looked in the past, Casesa mused, before the automobile became the dominant mode of transportation. "A hundred years ago the view would have been of horses. And now there were these devices that people were piloting at high speed." Casesa thought about the pace of change, and how the change would continue into the future. That's when he got his epiphany. Why *couldn't* cars be self-driving? Casesa thought.

Digging around for projections about the way a self-driving future might play out, Casesa came upon my paper. He read the analysis that projected just nine thousand autonomous taxicabs could provide Manhattan with superior mobility service at a tenth of the rate people currently paid for human-driven taxicabs. Casesa was a proud New Yorker who used taxis every day. "It knocked me off my chair," Casesa recalls. "Because it finally put an economic lens on this fantastic technology idea . . . The magnitude of the improvements that the paper was suggesting were far beyond anything I expected."

Casesa called me in July 2014, soon after he'd read the paper. "Your time has come," he told me, his tone near-exultant. Casesa and I had a lot of contact when I'd been running R&D and planning for General Motors, and he knew as well as anyone in the industry how long I'd worked to rationalize the transportation system—to address the system's inefficiencies and reform the wasteful use of energy, with vehicle electrification and hydrogen fuel cells. "In those cases, it was hard to see how to get from here to there," Casesa said, and he was right: When I'd been pushing them, the world wasn't yet ready for that technology, in part because it was difficult for consumers to obtain the hydrogen gas the fuel cells required to work—because your local gas station didn't top anyone's tank with the stuff.

But shared, autonomous mobility was different, Casesa observed. "The technology's finally caught up to your hypothesis," he said. "With self-driving cars, the savings are so profound that the investment is worth it . . . If you could make a self-driving car? You could change people's lives *tomorrow*."

Casesa began sending the paper to his colleagues and clients, who were among the most senior executives in the auto industry, teeing up the research by describing the changes it predicted as "an earthquake for the industry."

"This is not an *evolution* of your product," Casesa would tell them. "It's a *substitution* for your product."

At the time, automakers were coming around to the idea that autonomous technology was something they'd get to . . . eventually. We'll just keep adding safety equipment to our cars, they assumed, and the vehicles will get smarter and safer and more proactive and eventually you won't even have to *drive* the darn things. The automakers thought they would be able to ride this continuum from what Casesa calls "a dumb unconnected device to a connected, smart, self-driving device." But that was wrong, Casesa started to tell senior executives at the industry's most important companies. "The self-driving car is not what you make, only smarter. It's something completely different. It's a substitute—because now the hardware you make, the car, becomes a *service*." Before, mobility was a commodity in the form of individual cars that the companies sold. In the future, mobility was a *service* the companies would operate over time. The automakers would hold on to the vehicles they made and operate fleets of self-driving vehicles for themselves.

Overnight, the epiphany had an impact on the advice Casesa was giving to his clients. He began suggesting they prepare for a much different industry. Perhaps, he said, they should allocate capital *toward* certain new technologies, like the development of cheap and reliable LIDAR sensors, and *away* from old ones, like the catalytic converters that scrub toxic gases from the exhaust of gas-powered engines—which wouldn't be required in the emission-free electric vehicles that would dominate mobility-on-demand companies.

One of Casesa's clients that was most receptive to this thinking was Ford Motor Company, which had that spring just undergone a major leadership transition as former CEO Alan Mulally retired after nearly eight years leading the company, to be replaced by Mulally's chief operating officer, Mark Fields.

Casesa knew the company well. He'd followed it as an analyst for years. Now, as an investment banker, he maintained relationships with the company's highest ranks—including Fields and Executive Chairman William Clay Ford, Jr., commonly referred to as Bill. With them, Casesa began discussing the profound changes that were possible if this revolution played out. In the future, who was selling the car? Who was *buying* the car? How would Ford compete? Was Ford approaching a transition from selling single cars to hundreds of thousands of different people a year, and toward selling hundreds of thousands of cars a year to a handful of major fleet operators? Or operating the fleet itself? "The more we understood the ideas in [the] paper, the more interesting to us these new market opportunities became," Casesa said. Ford began to realize that a pivot toward a mobility company could represent a substantial growth opportunity—far beyond the company's current market.

It was Fields who first mentioned Casesa should go to work for the company the Ford executive now led as CEO. Working for Ford required moving to the Detroit area. Casesa loved New York. His wife had a job there. He had a son in high school he didn't want to uproot. But the opportunity intrigued him. While he considered the suggestion, Casesa accepted Fields's assignment to perform a strategic assessment of the company. He put together a small team and conducted a thorough, objective analysis. He was anxious about how his feedback would be received. Under Mulally, Ford had been regarded as one of the greatest turnaround stories in recent economic history. A *New York Times* article about Mulally's retirement compared the former CEO to one of the auto industry's best-regarded turnaround artists, Lee Iacocca, crediting Mulally with changing the company "from an industry laggard into one of the most successful carmakers in the world"—and noted that the week before Mulally announced his retirement, Ford reported its nineteenth consecutive profitable quarter. And now Casesa was going to criticize what Mulally had helped to create? He worried all he'd be doing was making enemies rather than making change. But it turned out Casesa had nothing to fear. Fields and the rest of the team didn't bristle at his criticism. Instead, they agreed with him. They could see things were poised to change in a big way.

"For me, a lightbulb went on," Casesa recalls. "That this company was not like some of the others in the auto industry that had this huge recovery, got a fresh start and were on top again—they didn't feel that way at all. They felt, if anything, a great sense of insecurity about what the future would be like. And their ability to compete in it. And they were very motivated to do something about it."

Another unusual thing about Ford was its leadership, in the form of Bill Ford, founder Henry Ford's great-grandson and the leader of the Ford family, which still retained a controlling interest in the $43 billion automaker. Bill Ford was an unusual figure compared to the stereotype of the Detroit car guy. He grew up in the sixties and went to Princeton, and he was acutely aware of the negative side effects of the invention his great-grandfather popularized. He'd long acknowledged that something had to change in the cities of the world, in speeches that sounded warnings about global gridlock and his belief that the planet didn't have enough room for more cars.

Maybe, Casesa figured, this job at Ford was an opportunity he shouldn't pass up. During a phone call with Bill Ford intended to explore the idea of Casesa joining the company, the family patriarch spoke emotionally about his commitment to transform the business, to do whatever it took to create something that would thrive in the auto industry's future landscape, even if that new entity looked very different from the current company. Bill Ford's open-mindedness about what that would entail, and his commitment to the transformation, impressed Casesa. "It was clear he defined his life's mission as making this company great for the next hundred years—and that he wanted help," Casesa recalls.

On the verge of accepting the offer and looking for someone to talk him out of it, Casesa called up Mulally, who had recently accepted a position as a new member of Google's board of directors. (Alphabet hadn't yet been created.) "Alan, I've been offered this job, it seems really interesting, I think I'm going to take it—but I want to talk to you first," Casesa said.

Mulally asked Casesa to read him the job description—for a new position, which Ford eventually called group vice president of global strategy, reporting directly to CEO Mark Fields and charged with developing the company's approach to autonomous cars and shared mobility, among other responsibilities. When Casesa was done reading out the job description, there was a long pause on the other end of the line. And then once Mulally began speaking, he didn't stop for forty-five minutes.

According to Casesa, Mulally said that when he started at Ford, in 2006, the job required some enormous tasks—but many of those tasks were obvious. Ford knew what it took to be a great auto company—the company just wasn't doing it. So Mulally did it. He restructured the company around a known business model. He made Ford into a great auto company again.

But *this* task of responding to the disruption the self-driving car represented—this was different, Mulally said. Sure, everybody knew what a great auto company was—but what was a great *mobility* company? Here was an opportunity to create, for the first time, an entirely new business template. To do something that had never been done before.

Mulally's perspective sealed the deal for Casesa. The former Ford CEO was as confident as Casesa that the industry was heading for a major disruption, one that would require the invention of a business that had never before existed. "I felt strongly the industry was at an inflection point," Casesa said. "Technology, connectivity, autonomy was being invented that would radically change the industry." Casesa felt like he'd been presented a historic opportunity to help participate in the creation of one of the world's first great mobility companies. "I have to do this," he told his wife. "If don't, I'll always regret it."

— — —

Ford announced Casesa was joining the company on February 17, 2015. The move would become the first major step by an American automaker to embrace the inevitability of an autonomous future—

as well as the first step in Ford's transition from an auto company to a *mobility* company.

Across the continent, Uber was making its own preparations for the future. "I think it starts with understanding that the world is going to go self-driving and autonomous," Kalanick would say later to journalist Biz Carson. "Because, well, a million fewer people are going to die a year. Traffic in all cities will be gone. Significantly reduced pollution and trillions of hours will be given back to people—quality of life goes way up. Once you go, 'All right, there's a lot of upsides there,' and you have folks like the folks in Mountain View, a few different companies working hard on this problem, this thing is going to happen.

"So if that's happening," Kalanick continued. "What would happen if [Uber] *weren't* a part of that future? If we *weren't* part of the autonomy thing? Then the future passes us by, basically, in a very expeditious and efficient way."

Soon after Drummond recused himself from attending Uber's board meetings, Kalanick assigned his chief product officer, Jeff Holden, to develop Uber's self-driving capability—to basically create the ride-sharing giant's own version of the Chauffeur project—and as quickly as possible.

Holden, whose drive is said to compare with Kalanick's own legendary focus, quickly identified the world's single greatest concentration of self-driving brainpower—at least, outside of Mountain View: Carnegie Mellon's National Robotics Engineering Center. Just to reiterate how tightly knit this community is, NREC (which everyone pronounces "EN-rek") was founded by Red Whittaker in 1996, with funding from NASA, and it also was the onetime employer of Chris Urmson and Bryan Salesky. It is located on the shores of the Allegheny River in Pittsburgh in a secure, industrial space that evokes even parts laboratory, robotics museum and mechanics shop. NREC is intended to work with American corporations to commercialize the technology developed by Carnegie Mellon's Robotics Institute. The place feels like something out of a Hollywood movie, the living embodiment of the lab in Pixar's

Big Hero 6, where one lab features CHIMP, an autonomous human-oid robot, and the next features robotic agricultural equipment designed for John Deere. Deeper into the facility, there's an area where visitors aren't allowed to enter, which presumably houses the secret projects that NREC conducts with the U.S. Department of Defense. With a staff of a hundred scientists and engineers and an annual operating budget of approximately $30 million, NREC had conducted its own share of pioneering autonomous-driving research. In fact, Salesky had explored an earlier deal in which Chauffeur would hire NREC's self-driving talent, but the idea didn't go anywhere because the NREC scientists didn't want to move to California and Google didn't want to split up the self-driving-car team.

John Bares had been working in and around Carnegie Mellon's robotics community since studying under Red Whittaker as a graduate student in the eighties. Bares, whom Kalanick would later describe as having an "aw-shucks feeling to him," was NREC's longest-serving director, having run the place for thirteen years, from 1997 to 2010, until he became frustrated by the unwillingness of NREC's corporate partners to commercialize the technology the engineering center developed. "We'd do a fantastic job engineering an advanced prototype, so-called throw it over the fence—and it would sit," Bares told the *New York Times.* "I just had the itch to do product." He left NREC and founded a start-up, Carnegie Robotics, which developed a land-mine-detecting robot that he sold to the U.S. Army, among other projects.

The first approach from Uber to Bares happened in November 2014 an email from Uber product and engineering executive Matt Sweeney asking whether Bares wanted to work on a "hard problem" at Uber. At first Bares thought it was a joke. "I'd never heard of Uber," he said, which was somewhat amazing even in 2014. To convince him of the seriousness of the offer, Kalanick flew out to Pittsburgh for a meeting. One goal then being discussed was Uber's hope to have 100,000 self-driving taxis on the road by 2020. Another: to be testing autonomous cars on public roads by August 2016.

Bares was a great pick to coordinate the staffing of Uber's version of Chauffeur because the thirteen years he led NREC meant he knew everyone who mattered, and he was respected in the community. Uber offered Bares a job late in 2014. But Bares turned down Uber's offer, apparently because the position would require him to move away from the Pittsburgh area. Holden and Kalanick persisted. Kalanick told Bares that he felt a pressing need to make automobiles more efficient, decreasing their environmental impact. The driverless future was coming, which meant Uber had to race to develop its own in-house autonomous capability before another company rolled out a model that would substantially undercut Uber's prices. Another offer featured a geographic compromise: It allowed Bares to stay in Pittsburgh, where Uber's self-driving-car team, known as the Advanced Technologies Group, would be located. Bares began discussing the opportunity with NREC engineers. When his colleagues heard the salaries that Uber was proposing, they convinced Bares to accept.

Bares joined Uber in January 2015. In the weeks that followed, the former robotics professor engaged in a furious rush of meetings and recruiting offers. The company put up an ad on a billboard outside CMU's computer science building that read, "We are looking for the best software engineers in Pittsburgh." But the primary target of Bares's recruiting was the institute he'd spent thirteen years running. In February 2015, Uber and Carnegie Mellon disclosed news of a joint project that would see the two entities working together to develop driverless-car technology. In retrospect, that agreement appears to have been a hedge. Even as the deal was being drafted, Bares was furiously recruiting to staff the Advanced Technologies Group, which soon would be housed in a 99,000-square-foot former restaurant-supply warehouse located less than a mile down the Allegheny from NREC. Bares was reportedly offering compensation packages that included signing bonuses in the hundreds of thousands of dollars and salaries at least double what the scientists and engineers had made at NREC. The researcher who had succeeded Bares in leading the center, Tony Stentz, accepted the offer. So did Bryan Salesky's former thesis adviser, Pete Rander. All told, forty NREC staff would leave—six principal investigators and thirty-four engineers. Working through Bares, Uber had essentially gutted the place. "I've never seen anything like it," marveled one Carnegie Mellon observer. "People have been complaining for years that no one understands how important this technology is. Then Uber came in and people were like, wow, this thing *is* real."

—— —— ——

Uber's mass hiring of NREC's self-driving talent created an enormous amount of discussion in the auto industry. It amounted to a high-stakes endorsement of Google's investment in Chauffeur. If Kalanick was right, if the market for on-demand mobility was enormous and did feature a significant first-mover advantage, then anyone with an interest in the space needed to make a serious play to participate—and fast. A *Wall Street Journal* article about the move, dated May 31, 2015, pinged around the upper echelons of Detroit-area executive suites. People marveled at the audacity of the maneuver—although it made perfect sense. *Of course* someone had picked off that dense concentration of self-driving talent. And *of course* it had been the hard-charging Uber, so quickly remaking the mobility world, that did it.

The coup set off a reaction in the overall industry. I think of what happened to the mobility space in 2015 and 2016 as a stampede, and this one occurred similarly to the way stampedes start in the wild—or, at least, in *The Lion King*. Visualize a herd of wildebeests grazing on an African savanna. One of the magnificent animals raises a head, sniffs the air, cocks an ear, frowns. She takes one ginger step forward. Then another. A little ways off, a different wildebeest stops its ruminating, raises *its* head and gazes at the first animal. "What does it know that we don't?" the second one thinks. It, too, steps forward. Next, two more of the beasts sense the movement of their herdmates. In this manner, four animals becomes eight, sixteen, thirty-two—and somehow, abruptly, there's a panic, until every member of the herd is sprinting down the savanna in the direction set out by that single, solitary animal, the one that first sniffed the disturbance.

Google is that first animal, thanks to the Firefly. Their single decision to design a vehicle operated almost entirely with software for full driverless operation, without a brake pedal, without a steering wheel, without even an accelerator, attracted crucial attention from within the mobility sector. Uber is the second, with Travis Kalanick so rattled by potential competition from Google and the "existential threat" of autonomous vehicles that he triggered a sequence of events that essentially disemboweled NREC. And you could make a good argument that coming a close third was Ford Motor Company, whose executive team was listening to John Casesa—and the warnings of its chairman, Bill Ford.

I'd been saying for years that self-driving technology was going to be the biggest thing to hit the auto industry since the invention of the automobile. Now, in 2015, people were starting to believe it. And why not? Uber would soon become more valuable than General Motors. Depending on the quarter, Google was the most valuable or second-most valuable company in the world. There were quarters that it could have bought GM outright with its cash reserves. The fact that the world's most valuable company and the world's most valuable privately owned company were both gunning hard to provide mobility on demand finally helped to convince people.

Soon, reporters were running stories describing "the industry upheaval." Renault-Nissan CEO Carlos Ghosn announced plans to sell ten new autonomous vehicles before 2020. Toyota's race-car-driving president, Akio Toyoda, had long opposed autonomous technology, from the company's motto, "Fun to drive," to promoting a corporate climate that reportedly "viewed the word 'autonomous' as taboo." Then Toyoda himself admitted to "a big change in my thinking," and announced a plan to invest a billion dollars to start a 200-researcher artificial intelligence lab in Silicon Valley designed to bolster the company's autonomous capability. Soon it was hiring some of the best robotics minds out there, including the former head of Google Robotics, James Kuffner. Toyota also hired all sixteen employees of MIT self-driving spin-off Jaybridge Robotics. The company promised to have vehicles driving themselves on highways by 2020.

The other thing that helped trigger the stampede in 2015 was the growing incorporation of self-driving technology by existing, established automakers. Tesla made its self-driving Autopilot software available to owners of Model S vehicles in October 2015. The project bore more than a passing resemblance to Google's discontinued highway-assist project, allowing the vehicle to drive itself once activated on a highway. Infiniti's Q50 featured "lane departure prevention" that would intervene if the technology sensed the driver was allowing the vehicle to wander out of its lane; as well, the CEO of Infiniti's corporate parent, Renault-Nissan Alliance, vowed to debut by 2020 technology that would allow its cars to navigate themselves through intersections. Daimler AG demonstrated a self-driving concept car that looked more like the sort of place one might take a meeting. Even mass-market vehicles like the Ford Focus were incorporating into their features such offerings as parallel-parking assist, which handled the reversing and gear-shifting if the driver pulled up alongside a free space and activated the feature. And in the name of safety, Volvo featured a pedestrian detection system that automatically braked the car if it sensed an imminent collision with a human being.

The biggest and most startling about-face of companies that hadn't been on board with development of self-driving technology came from the Detroit automaker to which I felt most personally tied: GM. The company had responded to Google's entreaties with an arrogance that had put off the self-driving-car team. But then CEO Mary Barra began listening to some new faces in GM's world headquarters, the Renaissance Center in downtown Detroit. The most influential among them at remaking the company's attitude toward autonomous cars was GM president Dan Ammann. "Lots of people say they love to drive," Ammann told *Fast Company*, speaking in the context of the 86 percent of Americans who commute to work by car. "But I haven't met anyone yet who says they love their commute." Taking into account the rise of ride sharing, millennial ambivalence to vehicle ownership and the increasing safety of autonomous technology, Amman's leadership prompted the team to realize that changes in mobility were coming—and that they could destroy GM's business.

The company's first steps toward embracing the new technology were painful. In summer 2015, GM staged a press event at its proving grounds in rural Michigan designed to show off its own self-driving technology, to prove that they were leaders. But in a subsequent feature in *Bloomberg Businessweek*, the awkwardness of the company's position was evident. GM product development chief Mark Reuss demoed the Cadillac Super Cruise's highway driver's assist product for a reporter, and as he did, he seemed to be a bundle of nerves—"on edge" and "forcing a nervous laugh" as the car accelerated up to 70 mph. The story portrayed Reuss as this wide-eyed kid riding in a self-driving car for the first time. And this was in a controlled environment on GM's own proving grounds. Reuss was anxiety-ridden demoing stuff that the guys at Carnegie Mellon had been able to achieve eight years before on a university research budget.

Luckily, General Motors CEO Mary Barra had begun sending executives to Silicon Valley on scouting expeditions that aimed to identify possible acquisitions. Ammann's influence teed up some savvy moves—although, at this late stage in the game, GM paid hundreds of million more in deals than it would have, had it gotten on this thing just two or even three years before. On January 4, 2016, the company announced it had invested $500 million in the ride-hailing company Lyft. Later the same month it launched a Zipcar-like ride-sharing company, Maven. And on March 11, 2016, GM announced its intention to spend $581 million, including $300 million in cash, on a forty-employee Silicon Valley start-up called Cruise Automation, in a bid to obtain the brainpower that could pull it into an autonomous future.

But the most startling thing about the transition at GM was the December 2015 essay that CEO Mary Barra published on LinkedIn. "I believe we'll see more change in the automotive industry in the next five to ten years than in the past fifty," she said. "I have committed that we will lead the transformation of our industry." The essay's headline proclaimed 2016 as "the year Detroit takes on Silicon Valley."

I agreed with Barra about the next five to ten years. But GM was at least five years behind. It wasn't going to *lead* anything. GM had every opportunity to do what Google did in 2009—and it had even more of an opportunity to do what Uber did in 2015, which is to say, hire some of the best robotics minds in the world and point them at the rapid development of self-driving cars. GM, after all, had a long-standing relationship with Carnegie Mellon. Which raises the question: Why *didn't* they? In fact, why did it take Detroit, as a whole, so long to realize the importance of this stuff?

One of my good friends within Chauffeur is Adam Frost, who worked for Ford for eighteen years, most recently as a chief engineer. He was one of the early industry figures to catch on to how transformative this would be. At Ford, in the Australian market where he worked, one of the things that made Frost proudest was the way his advocacy convinced the company to make head-protecting airbags standard on low-cost cars. The Ford veteran was attracted to Google's project because he saw it as another step toward safer cars, one that saved lives by a couple of orders of magnitude more than the head-protecting airbag innovation. "The airbags saved maybe a couple of hundred lives a year," Frost reckons. Self-driving cars, as I've argued, could save more than a million.

"I think we were asleep at the wheel," Frost admits, describing the automakers' attitude toward the self-driving-car project. "If Google had been doing the testing in Detroit, maybe the industry would have caught on earlier." One of the problems, Frost points out, is the nature of the testing process. Developing a self-driving car requires a lot of testing in public, millions of miles of it, as the engineers attempt to ensure the vehicle reacts appropriately to every imaginable scenario. But testing in public is anathema to the closed-door approach that Detroit prefers, which usually occurs at remote, ultra-private proving grounds.

Then there's the innovator's dilemma. It's really hard for big companies to disrupt themselves. The examples of companies that have done it well are rare. "One of Google's gifts was that they didn't have a product in this space," says Chris Urmson. "They didn't have to worry about how it fit into an existing portfolio. They could just focus on the next thing."

Casesa agrees with that line of thinking. "Google wasn't in that business—they didn't have anything to undo," he said. Then he names a third reason Detroit was so slow in getting on board. "People say auto companies don't take enough risk. Well, in a company like General Motors or Ford, there are places where you want to take risks and places you don't. Designing a brake system—you don't want to take risks. Building a new plant—you don't want to take risks. Launching a new product—you don't want to take risks because it's got to be safe, it's got to be all of that stuff. So the very conservative, risk-averse culture makes a lot of sense when you're designing cars, when you're building plants, when you're creating the safety-critical products. But when you're doing a prototype or trying to test a new business model? That culture is a barrier to success."

That car companies make hardware is another factor. They design steering wheels and headlights and door handles, and they're really good at getting it all in the same building at the same time in order to assemble an automobile that's going to work in hot and cold and night and day for hundreds of thousands of miles. But the self-driving question is essentially a software and mapping problem. It requires writing lots of computer code, which isn't a car company's strength. The automakers looked at the Google self-driving-car project, and they foresaw a future in which the automobile became the latest example in a business trend that had already affected two other software-heavy devices, personal computers and smartphones. Early in the onset of the personal-computer revolution it had been the hardware makers who had all the power, who had been the most valuable brands—the Texas Instruments, the Commodores, the Hewlett-Packards. Then it turned out that the real differentiator when it came to a computer was the software that was on the device—and the power migrated to the OS makers, the Microsofts and Apples, until the hardware makers were little more than afterthoughts. The phenomenon is even more apparent with smartphones—very few people who use Apple's iPhone know or care who actually assembles the device. In the case of the automakers, why would they do anything to usher in a technology poised to make them just as irrelevant?

Finally, as I've pointed out, getting to full autonomy, the kind that allows you to think it's a good idea to make a car without a steering wheel or brakes, requires really good maps—a high-resolution 3-D scan of every road the self-driving car is going to ride, ever. Which, in the early days, would have seemed insurmountable to nearly any company or researcher that *wasn't* Google. Turn that around: Google is one of the only companies where the executive suite wouldn't have batted an eye if you told them you were working on a solution to the self-driving-car problem, and that solution required the high-res mapping of every single road in the United States, and probably one day, the world.

In fact, the reason the auto industry was slow was that they didn't have a bone-deep understanding of digital technology, or the full capability of computers and big data. They didn't understand cutting-edge communications technology. They would never have developed this technology on public roads. And because they're primarily car guys and bean counters, as opposed to mobility types, they tended to believe their business was about manufacturing and selling cars—when actually, the real value creation mechanism of the automobile industry amounts to helping people get from one place to another.

Chapter Eleven

DRIVING OPPORTUNITY

Exaggeration is a billion times worse than understatement.

——FRANK L. VISCO

While the automakers and their parts suppliers were stampeding into the personal-mobility market, big things also were happening inside Chauffeur. The events reflected the way the project was leaving its start-up phase. The next step involved commercializing what the team had created. In September 2015, the self-driving-car team learned that Google had hired auto executive John Krafcik to take on a senior leadership role. The start of the following month marked the formal beginning of the Alphabet Inc. holding company created by Larry Page and Sergey Brin to manage the assets formerly operated under the Google flagship brand. Both moves foretold the spinning out of the Chauffeur team into its own new company, to be known as Waymo, the following year. (The name is a nod to the company's mandate, to provide "a new way forward in mobility.")

Later that same month, the Chauffeur engineers staged one of the most audacious demonstrations yet of their technology, in Austin, Texas. A Firefly vehicle transported a man named Steve Mahan around the city without a safety rider or, indeed, anyone else in the vehicle but Mahan himself. Firefly had a stop button within easy reach, to allow riders to halt the vehicle in case of emergency. Mahan wouldn't have been able to spy any troubling circumstances, however, because he was blind. Such was the Chauffeur team's confidence in its technology.

The poisonous dynamic between Urmson and Levandowski was as bad as ever. Levandowski continued to develop LIDAR technology within Chauffeur, working on various innovations that aimed to reduce the sensor's cost and improve its capability. At the same time, Levandowski is alleged from 2013 through 2015 to have been involved in a pair of competing side businesses known at various times as Odin Wave and Tyto LIDAR, which also had been developing LIDAR technology. Odin Wave first came to Google's attention in July 2013, according to the company's Arbitration Demand with Levandowski, after one of Chauffeur's parts suppliers, a hardware vendor, told a Google employee that a company named Odin Wave had requested the vendor custom-fabricate a part that was remarkably similar to another part that Google used in its proprietary LIDAR. At the time, Google assigned two employees to investigate Odin Wave. They learned that the company was located in a building that Levandowski owned. According to the Arbitration Demand, "Levandowski was questioned about his affiliation with Odin Wave in mid-2013 but denied having any ownership interest." Odin Wave merged in February 2014 with Tyto LIDAR, which Google alleges listed a friend of Levandowski as its manager. "Google . . . believes that Levandowski had some involvement with Odin Wave/ Tyto since at least 2013 at the same time that he was working on Google's development of LIDAR sensor modules," the Arbitration Demand alleges. In the spring of 2015, apparently looking to neutralize a possible competitor, Chauffeur convened a committee to investigate the company now called Tyto LIDAR without realizing that Levandowski had any involvement with it. In fact, according to the arbitration document, "Levandowski participated in Google's investigation into Tyto's products and business, including at least one site visit to Tyto headquarters . . . [H]e was privy to Google's impressions of Tyto's products and process, including Google's confidential opinion of Tyto's technology and the viability of Tyto's business. Throughout this process, Levandowski never disclosed a relationship with Tyto and its employees." The committee ultimately opted not to move on the deal.

The decision coincided with a new and final phase of Levandowski's involvement with Chauffeur. Previously, he had worked as a relatively functional part of the Chauffeur team, possibly because his job allowed him access to the knowledge base the company was developing around LIDAR technology. Then, when Google opted not to acquire Odin Wave, Levandowski checked out, according to coworkers, doing the bare minimum required to remain employed with Chauffeur until the bonus plan vested. That summer, after the creation of Uber's Advanced Technologies Group, Levandowski is alleged to have discussed with members of the Chauffeur LIDAR team the possibility of leaving Google and joining Uber as a unit, telling his longtime colleague Pierre-Yves Droz that it "would be nice to create a new self-driving-car start-up." If they did, Levandowski said he thought that Uber might be interested in acquiring it. Word of Levandowski's recruiting reached Urmson, who was concerned enough to send an email to Google HR. "We need to fire Anthony Levandowski," Urmson wrote. "I have just heard today from two different sources that Anthony is approaching members of their team attempting to set up a package deal of people that he could sell en masse to Uber." Levandowski has not commented publicly on these accusations, nor did he respond to questions about these activities in court proceedings between Uber and Waymo.

—　—　—

Consider that during this same period Google and the Chauffeur team of engineers were moving toward negotiating Chauffeur's valuation with a view to paying out the bonus plan at the end of the year. The size of the bonus payout was tied directly to Chauffeur's valuation as decided by Google. The valuation would increase if Chauffeur had an agreement with a major automaker. Pairing up with such a company would help get the technology into vehicles used by consumers in the real world—into hundreds of vehicles, with the understanding that those hundreds would lead to thousands, and tens of thousands, and so on, until the technology was proliferating across the globe. So in the spring of 2015, as Uber was hiring the NREC researchers from Carnegie Mellon, Chauffeur began talks with Ford Motor Company.

Ford was as good a fit as any in the auto industry, I thought, thanks to board chair Bill Ford's progressive attitudes about the auto industry's global effects, and former Ford CEO Alan Mulally's position on the Google board. Plus, a big deal with Ford would have tied Chauffeur to Ford's impressive heritage.

On November 17, Urmson staged a meeting for the rest of the Chauffeur team to announce the valuation. The number was $4.5 billion—a remarkable figure considering that Google had reportedly calculated that it had spent $1.1 billion to develop the business. That might sound like a lot, but developing new technology can be expensive. (GM spent about the same amount developing fuel cells during my eleven years as their vice president of R&D.) So in just seven years, Urmson had led day-to-day operations on an effort that created $3.4 billion more value than Google had spent on the project. Not a bad result. For that, Urmson hoped he would win the CEO position. Instead, we learned that John Krafcik would become the new entity's CEO. The plan was for Urmson to become the chief technology officer. I sympathized with Urmson but agreed with the Google leadership that the project required high-level management experience in the auto industry—which Krafcik had in spades.

The Ford deal that Chauffeur was negotiating helped to form this thinking. For proprietary reasons, I cannot divulge much about the negotiations with Ford. But consider that Ford CEO Mark Fields and his chief product development and technology officer, Raj Nair, were accustomed to some of the most intense bargaining sessions on the planet. They negotiated with the UAW. They negotiated with their suppliers. They negotiated with their dealer networks, and they negotiated with lawmakers over regulations and tax incentives when deciding where to situate a new factory or product line. To be a senior executive at a Detroit automaker meant you were, quite simply, one of the best negotiators on the planet. The Ford team approached the negotiations like a standard zero-sum auto-industry deal, by which I mean, any concession that Ford won was a loss for Google, and vice versa, and the end object was to get out of the bargaining process making as few concessions as possible. In contrast, what was needed was a deal in which both sides won simultaneously.

The media reports about the deal between Ford and Google focused on the cultural disconnect between the two companies, which was indeed formidable. That December, Fields traveled to Mountain View for a series of meetings that included a meal with Brin and Page. Before it happened, Ford security staff showed up at the Google campus to examine the location, apparently as a kind of risk assessment, which raised a lot of eyebrows within Chauffeur and served to underscore just how different the two companies were. For one, Fields and his lieutenants arrived in a caravan of massive Lincoln Navigators to discuss a deal that excited many in Mountain View for the way it was supposed to bring the auto industry closer to a more sustainable existence. Meanwhile, Sergey Brin is a billionaire who rides his bike to work; when he absolutely has to drive, he uses a Tesla. The differences were greater than that, however. As an executive, running a hundred-year-old car company selling a broad portfolio of vehicles around the world, Fields led his life developing and executing plans on schedule. He was driven by timelines and process and had very little time left to think creatively. Brin and Page, by contrast, are two of the most intelligent people on the planet, as well as two of the most nimble in terms of creatively generating fresh strategies to navigate just-invented industries.

While the cultural divide helped doom the deal, it was not the only, nor the most significant, issue. Getting an agreement done was important to both sides. Chauffeur needed to get its technology into vehicles on the scale that was necessary to transform the way global society approached personal mobility. Ford could have done that. And it was important to Ford to the extent that it would link the Detroit company with the leader in the autonomous space.

But the two sides just couldn't communicate the way they needed to. The intended deal included a long-term agreement that could have resulted in the companies collaborating to get hundreds of thousands of autonomous vehicles in the marketplace. Creating a new vehicle that could scale to those numbers required an investment in the billions, and Google wasn't ready to commit to those kinds of numbers.

As the deal was being negotiated, Google leadership realized they needed someone to lead Chauffeur to commercialization who had experience in the way things were done in the auto industry. That person was Krafcik.

"If Silicon Valley and Detroit had a baby," went one industry report on the new hire, "it would be John Krafcik." I liked that. Krafcik's hiring struck me as a prescient recognition on the part of Google's leadership that the old Silicon Valley versus Detroit model was broken, and that to get this deal done, to actually trigger change on the scale that we'd all been discussing for years, required a new relationship between the auto industry and the technology sector. Something a lot closer to partnership.

Krafcik and I knew each other by reputation and hit it off from the start. Fifty-four at the time, and with graying hair, Krafcik came off as a lot younger, in part thanks to his runner's physique, but mostly because he's quick with a smile and approaches everything with a youthful enthusiasm. He comes from American car-guy heritage, having grown up the youngest child in a family of eight children in which the various siblings battled for first dibs on *Car and Driver* magazine. Special one-on-one time between Krafcik and his father involved tinkering with the innards of the family's '66 Oldsmobile F85, replacing spark plugs or changing the oil. In fact, his father's mechanical ability was such that he built, as hobby projects, two full-scale airplanes. One of Krafcik's brothers owned a series of Corvettes. Out of high school, Krafcik studied mechanical engineering at Stanford, and soon after graduating in 1983, he ended up at the New United Motor Manufacturing, Inc., factory in Fremont, California, a joint operation of Toyota and General Motors. There, the Toyota Production System was applied to the assembly of a new class of smaller, front-wheel-drive American vehicles, such as the Chevrolet Nova.

The Toyota Production System is better known today as "lean production," and is near-universally lauded for its ability to improve quality and spur productivity while minimizing waste. Krafcik, who would go on to become one of the world's experts in the approach, actually coined the term "lean production" in a 1988 article he wrote for the *Sloan Management Review*. After his work at NUMMI, Krafcik attended MIT's Sloan School of Management, where he got a job working for James Womack, the director of MIT's International Motor Vehicle Program. Womack would in 1990 write the definitive book on Toyota's production methods, adapting Krafcik's term for the subtitle: *The Machine That Changed the World: The Story of Lean Production*. On the American-style mass production lines, the emphasis was on keeping things going at all costs, regardless of whether a worker made a mistake, then hoping you caught all the mistakes in inspection and repair areas at the end. The lean manufacturing approach, meanwhile, amounted to doing everything right the first time, empowering any worker on a line to be able to halt production whenever a defect or flaw was spied.

After MIT, Krafcik joined the Ford Motor Company and stayed through the nineties, eventually becoming chief engineer of the Ford Expedition/Lincoln Navigator SUV product lines. Krafcik's then-wife was a doctor, and he envied her life's work for its ability to improve people's lives. Krafcik found a similar purpose when he became a champion within Ford for incorporating the latest safety features as standard products. This was the mid- to late nineties, the advent of side-impact and side-curtain airbags, which would deploy to prevent injury in the event of a rollover. Krafcik's advocacy helped to make them standard equipment on the Lincoln Navigator. It rankled him that he couldn't get them to become standard on the lower-priced Ford Expedition. An inability to convince the company's management to incorporate other safety features as standard was a factor in his leaving Ford in 2004. He landed at Hyundai, the Korean automaker, when it was still considered an inconsequential part of the American import market. He stayed for ten years, rising to become CEO and president of the company's North American operations. Before Krafcik's arrival, Hyundai had already begun to turn things around by providing the best warranty in the business. Krafcik helped by encouraging them to become among the most safety conscious, too, as well as spurring marketing that included Super Bowl ads and a pledge, during the 2008 recession, to accept the return of a vehicle if the owner lost a job. The unconventional tactics worked. Under Krafcik, Hyundai North America's annual sales climbed 75 percent, from 400,000 to 700,000, during a period that saw the industry's average increase at 19 percent.

While at Hyundai, Krafcik became an obsessive tracker of NHTSA's Fatality Analysis Reporting System (FARS) database that listed the circumstances behind each death that happened in an auto crash. The deaths he encountered there encouraged him to fight for more safety equipment in cars, such as lane keeping, forward-collision warning and automatic emergency braking. What he came to learn as he monitored FARS was the culpability of the human drivers in most crashes. "One of the realizations I had at Hyundai was, [safety features] didn't make as big a difference as we thought," Krafcik says. "Because more than ninety percent of accidents are caused by humans."

Which was why the invention of Firefly, a vehicle without a steering wheel or other controls, blew his mind. "Most people in the industry saw it and shook their heads and said, not going to happen—not in my lifetime," Krafcik recalls. "I saw it and said, 'Holy cow, it's amazing—*that's* where we have to go.'"

It was while he was at his next job, the online pricing site True-Car, that Krafcik first began talking with the recruiters who would bring him on to lead Waymo. During one of Krafcik's visits to Mountain View, Sergey Brin dropped by and asked whether he had time for a chat. Krafcik hopped into Brin's Tesla as the Google founder drove to pick up his kids from a nearby day camp. Brin pointed to various aspects of the vehicle and asked what they might cost to manufacture. They talked about what Krafcik thought of each of the major automakers as potential partners for Chauffeur. Krafcik in turn asked about the various members of the team, and it struck him that, besides Adam Frost, very few had any experience within the auto industry.

For his interview with Larry Page, Krafcik described what was known at Ford as the "occasional use imperative"—the idea that people buy cars not for their *primary* purpose, but for their unusual, occasional purpose. Take sport-utility vehicles, Krafcik said. Many can accommodate eight passengers, with all-wheel-drive powertrains able to tow ten thousand pounds. Most drivers use such capabilities, at most, once a year. For most trips we need a one- or two-passenger car to get to work and back. And yet those occasional uses stand as one of the primary drivers of vehicle sales. That sort of waste is certain to have upset any expert in lean production. "One of the things I'm most excited about this future that we can craft together is, we can replace this inefficient car stock," Krafcik told Page. "Which has so much more capability than we need, with cars that are suited just to the purpose."

In the future, Krafcik said, probably 75 percent of the cars on the road should be sized for just one or two people. And then maybe 20 percent will hold five or six people. The remaining 5 percent will be the behemoths that are most of what we see on the road today.

Page must have liked what he heard because Krafcik progressed on to his final interview, which required flying up to Seattle. Former Ford CEO and current Alphabet board member Alan Mulally picked Krafcik up at the airport on a Saturday in his Ford Taurus sedan and took him to a nearby restaurant for a bite to eat. The two men discussed the best way to bring the technology to market, including such possibilities as deploying it for the trucking industry, as a transportation-as-a-service business that provided rides to consumers, or licensing the technology to automobile manufacturers while retaining control of the sensors and software.

In my estimate, Krafcik's hiring to lead Waymo was a remarkable moment—one that represented a significant epiphany on the part of

Larry Page and Sergey Brin, and the rest of the Google leadership. For years, the Chauffeur team had not been troubled by its absence of auto industry experience. In many instances, ignorance of the way Detroit did things was considered an asset. But Krafcik was an auto industry guy, and his installment as the team's leader was the first acknowledgment to my mind that maybe this quest didn't have to pit Silicon Valley against Detroit. That each of the two sides had expertise to contribute toward solving the auto industry's waste issue. Krafcik's hiring was a savvy concession on the part of Google's founders that maybe, just maybe, they needed Detroit.

The new attitude toward collaboration, however, did not mean that the Mountain View executives were going to agree to a terrible deal with Ford. Sometime after Krafcik arrived in Mountain View he asked me what I thought of the terms. I told him I didn't think there was anything in it for Google. Ford appeared to see it more as an opportunity to sell more cars, than a chance to collaborate with the technological leader. That was short-sighted. And so the agreement was pared steadily back until it became, basically, a straight-up vehicle purchase. By the time the term sheet reached Page and Brin, there wasn't anything in it for Chauffeur.

They were right to walk away. The mere fact of a deal would have represented a significant boost to Ford's share price. And yet, Ford's negotiators were focused on nickel-and-diming. Another misplay was the Ford team's leak to media about the negotiations that December. "Ford is leaking the news to try to force your hand," I said to Krafcik once the news broke. But Chauffeur wouldn't play it like that.

Of course major companies often attempt to collaborate strategically and fall short. What happened with Google and Ford occurs all the time. But the dissolution of the deal would hurt Ford CEO Mark Fields a lot more than it affected Chauffeur, which simply went on to negotiate a better deal with Fiat Chrysler. Meanwhile, Fields's inability to get an agreement appeared to be among the factors that led to Ford's board of directors pushing him out the following year.

Through the course of 2015, Levandowski developed a friendship with Uber CEO Travis Kalanick, who had sought out Levandowski as someone who may have been able to provide him with some context on the Chauffeur project. The two technologists would walk miles around San Francisco, favoring a route from the city's Ferry Building all the way along the north shore to the Golden Gate Bridge, logging thousands of steps on Kalanick's pedometer and discussing ideas about the developing mobility ecosystem.

In the meantime, the day that the engineers had long anticipated, the vesting of the Chauffeur Bonus Plan, had arrived, and the payout stood to make them wealthy people. The bonus payouts were to be paid in a forty-sixty split, with the first to be disbursed on December 31, 2015. Levandowski was due to get $50,617,800 as his first payment—reportedly among the largest bonuses Google had ever paid, although in his deposition Larry Page took pains to point out that it was less an annual bonus and more like the sort of equity compensation provided to members of a start-up. Before he was actually paid the bonus, however, on December 11, 2015, Levandowski allegedly downloaded from Google servers approximately fourteen thousand technical documents related to Chauffeur's self-driving technology. He would later insist to Travis Kalanick that he needed the documents to help him work from home; he held on to them afterward, he would say, because he thought they might form a kind of collateral to ensure that he was actually paid his share of the Chauffeur bonus plan. Asked directly about the matter later in court proceedings, Levandowski did what's known in the States as "took the Fifth," declining to answer the question as is his right under the U. S. constitution in situations where providing a response could incriminate the respondent.

That same month of December 2015, Levandowski's discussions with Uber took on a new urgency. With the certainty of receiving his bonus, he apparently felt more able to consider his next steps. Kalanick and Levandowski had one meeting at Uber's San Francisco headquarters and then a second immediately after the first payout of Levandowski's bonus, on the weekend of January 2, 2016.

A week later, Levandowski sent Larry Page an email, which Page forwarded on to Krafcik and Brin. "L.," Levandowski began. "Happy new year, sorry for long-ish email . . . Chauffeur is broken. We're loosing [sic] our tech advantage fast . . . we should deploy the first 1,000 cars asap. I don't understand why we are not doing that. Part of our team seems to be afraid to ship. Deploying now would show where our system is not working and push the team to fix it asap instead of inventing new things which I think are not needed . . . Time to market is more important now than ever . . . I also don't understand why we aren't putting our tech on as many cars as possible (e.g., consumers or Lyft drivers, for example) to roll our [sic] our tech much faster and for less money."

There was a lot more ranting, besides. All of it was classic Levandowski. For years he'd undermined Urmson's leadership by exploiting his own direct connection to Page. Now he was doing the same thing to Krafcik. One of the most interesting things about the email was its subject heading: "'Team Mac' urgently needed." The "Team Mac" was an apparent reference to the secret set of software whizzes inside Apple deployed in the early eighties to build the Macintosh computer. Team Mac had competed against another Apple team then working on a more expensive machine, the Lisa. The subject heading was Levandowski proposing a second self-driving-car team within the Alphabet umbrella, one presumably led by Levandowski himself, that would compete with Chauffeur.

At some point that January, Levandowski spoke with Larry Page and floated the idea that the younger engineer might work on a self-driving truck as a side business. The matter came up as Levandowski was complaining about his Chauffeur colleagues. Levandowski said he was sick of the Chauffeur employees who didn't like him. And then, almost as though he was musing aloud, he said something like, "Why don't I just go do a company that does trucking? And everything will be fine."

Page reacted by pointing out that such a move could represent a violation of Levandowski's employment contract. "I told him very, very clearly that I thought that was highly competitive and not a good idea," Page recalled. "I'm like, No, that's not fine. Like, that's the same thing as what you're doing here. I mean, you can do that, but we are not going to be happy."

Meanwhile, Levandowski, still a Chauffeur employee, was working hard to convince fellow team members to join him in moving over to Uber. On January 20, Uber's head of the Advanced Technologies Group, John Bares, under pressure to get a self-driving car tested on public roads by Uber's deadline of August, and to meet a milestone of 100,000 autonomous vehicles by 2020, sent an email to the senior Uber leadership, including Travis Kalanick, Emil Michael and Jeff Holden, that described the value that Levandowski's team could bring to the company. "The nexus of this group represents a couple of very significant things to our AV effort," Bares wrote, using an acronym for "autonomous vehicles." "First, Anthony and his close team have developed several generations of mid and long range lasers that we now believe are critical to AV autonomy . . . We have yet to find anyone else in the world with this know-how. Our next best choice is to build the team internally and we can do that, but probably with a 2–4 year lag on what these guys can do.

"Second," Bares continued, "just rubbing shoulders with this team and having them advise us all over AV has a decent chance of saving Uber at least a year off of the race to large scale AV deployment . . . My point is that there is more value here (considerable) than 25 disparate engineers that we would pick up from 25 different places."

Three days later, Salesky sent Levandowski a note suggesting some new roles he might consider pursuing within the team. Two days after that, Levandowski had a phone call with Kalanick. And on January 27, 2016, Levandowski resigned from Chauffeur by sending human resources an email with the subject heading "It's time for me to move on."

Later the same day, Levandowski wrote an email to Larry Page that pointed to "too much BS" with Urmson, John Krafcik and Bryan

Salesky. Days after Levandowski's departure, in a move that shows just how small the world of star self-driving technologists is, Uber's John Bares called Chauffeur's Bryan Salesky. The pair knew each other from their NREC days. Bares was calling for a reference on Levandowski, and noted during the conversation such terms as "backstabbing," "bombing Google—he will do it to you" and "whispering upstream and not team playing." Within weeks, people within Alphabet were hearing that Levandowski was working on a trucking company—a rumor proven true when a company called Otto released a video that May of an eighteen-wheeler driving autonomously along a highway. Uber announced in August that it was purchasing Otto, then just six months old, for a reported $680 million. (The payout from Uber to Otto was conditional on Levandowski and others achieving an aggressive series of milestones.)

Each of the major automakers made similar moves to join the mobility ecosystem. Jim Hackett left Ford's board of directors to head up a new arm of the automaker, the Smart Mobility program. There was GM's deal to purchase Cruise. And in May, Google and Fiat Chrysler announced a deal to develop one hundred Chrysler Pacifica hybrid minivans, another step toward commercialization.

— — —

A tremendous amount of money and time had been invested in the mobility disruption, and yet, its execution was by no means a certainty. The auto industry's pursuit of big autonomous-vehicle deals meant that Detroit was now working in collaboration with Silicon Valley to bring about the future. Although Ford and Google couldn't find a way to cooperate in developing autonomous vehicles, the two companies could, that April, join forces with Uber, Lyft and Volvo, among others, to create a trade group, the Self-Driving Coalition for Safer Streets, to lobby the federal government to create a hospitable regulatory environment for the testing and eventual rollout of the new class of transportation.

Detroit, which previously had lobbied to slow down the development of driverless cars, had now joined Google in working to hasten their arrival—a position shared by others who testified alongside the Self-Driving Coalition at NHTSA hearings, including Mothers Against Drunk Driving and groups advocating for the disabled.

But not everyone was so enthusiastic about the new technology. And who could blame them? If my projections were correct, it stood to transform entire industries. That June I spoke at a conference staged by one of the world's biggest producers of oil exploration equipment. It was held at the company's research center, and the room included some of the industry's top-ranked executives. "This disruption represents the first opportunity in more than a century to remove the automobile from the energy and environmental debate," I told them.

America's vehicles consumed about 180 billion gallons of fuel a year—which makes up about half of total U.S. oil consumption. But all of that was about to change. As I went through my presentation I could see the audience taking in the implications of what I said. Foreheads knitted. Mouths turned to frowns. The concern of the oilmen and -women was palpable as they considered the future I predicted, in which 80 percent of U.S. miles traveled happened in driverless vehicles powered by battery-electric and other alternative-propulsion technologies.

All of it meant that demand for oil to support America's transportation needs could fall to less than a fifth of what it is today. A similar phenomenon will happen all over the world as gas-powered cars give way to a constellation of propulsion options. In fact, the oil industry could be looking at a future not unlike what the coal industry has been experiencing over the past several decades.

A worldwide reduction in demand for oil and gas neuters OPEC, reduces the power of Russia and Venezuela, and creates some major problems for the regions of America that depend on income from the oil industry, such as Texas. It may also create more of a threat from terrorism, since there's a potential for greater numbers of disaffected youth in a much less wealthy Mideast. In fact, the vested interest of the oil industry in the historical automobile system could represent the biggest threat to the mobility disruption. After all, oil producers realize enormous profits from the automobile's thirst for gasoline. Consider that in 2008, the same year that GM lost $31 billion, thanks in part due to shifts in consumer demand because of high oil prices, Exxon made $45.22 billion in annual profit—the largest in U.S. history.

How will well-monied corporations respond to the threat of mobility-on-demand services? My friend and colleague Robbie Diamond has played out some of these scenarios in his head. Diamond runs SAFE, a Washington, D.C., nongovernmental organization, the initials of which stand for Securing America's Future Energy. His organization's advocacy helped to ensure that Congress passed the Energy Independence and Security Act, the first legislative change to corporate average fuel economy standards since 1975, which President George W. Bush signed into law in 2007. Diamond was able to do that because he portrayed America's fuel efficiency as a *security* problem in addition to an environmental problem. Even so, the auto and oil industry fought his efforts with a well-orchestrated lobbying campaign. That fight taught him the way well-funded special interest groups in Washington are able to derail initiatives that otherwise make perfect sense for the populace at large. "In Washington," Diamond says, "it's a lot easier to *stop* something than it is to make something happen."

As part of Diamond's advocacy for a more rational, less oil-dependent transportation system, he's rooting for the mobility disruption. Diamond also wants it to happen for a personal reason. He has a ten-year-old daughter whose legs are two different sizes, and they don't bend like they should. He wants autonomy to happen for her, and the 5 million Americans whose disabilities prevent them from driving, including 1.5 million blind Americans.

"Right now we're totally dependent on oil in the transportation sector," Diamond says. "Oil is the most important commodity that makes the world go round. But if the mobility disruption happens, that'll no longer be the case."

Another reason we need this to happen is that American roads are growing more deadly—despite our best efforts. Fatal-traffic-crash data for the calendar year 2016 report 37,461 deaths on U.S. roads. That represents an increase of 5.6 percent compared to the year before. Until recently, the number of annual auto-crash fatalities had shown a five-decade decline as safety technology improved. Lots of reasons exist to explain the recent increases. One major factor is the temptation that results from toting along a smartphone pretty much everywhere. It's just too easy to glance at a screen every time a notification comes through, regardless of whether you're driving.

Hence the hope that door-to-door autonomy can lead to the complete elimination of such fatalities, thanks to the invention of software that amounts to the world's best driver. Consider what happened the first time that a vehicle equipped with the Chauffeur software bore responsibility for an accident. It happened on February 14, 2016, with Urmson's team having logged roughly seven years and 1.45 million miles of testing on public roads before that without any at-fault incidents—an extraordinary safety record.

The accident happened at the intersection of El Camino Real and Castro in Mountain View, not far from the Google headquarters. At that point, El Camino Real features three lanes of traffic in each direction separated by a boulevard strip. One tricky thing about El Camino is that the rightmost lanes, closest to the curb, can be wide enough to accommodate two cars alongside each other. On the day in question, heading eastbound on El Camino toward Castro, one of Chauffeur's self-driving Lexus vehicles got into one of those wide right-side lanes, moved right and activated its turn signal to prepare to turn northbound on Castro. Blocking the progress of the Lexus was a pile of sandbags surrounding a storm drain.

The Lexus couldn't proceed forward because of the sandbags, and it couldn't go around the obstacle because cars had come up along its left side in the same lane. So the vehicle stopped until the light turned green and the cars moved forward. Just as the Lexus was angling back deeper to the center of the lane to get around the sandbags, a transit bus came up behind it. The software managing the Lexus figured the bus would stop—after all, the Lexus clearly was angling deeper into the lane. The test driver behind the steering wheel of the autonomous Lexus figured the bus would stop, as well.

But the bus didn't stop, and the Lexus scraped up against the side of it. No one was hurt, as the Lexus was going only about 2 mph. But the Lexus dented its left front fender and damaged both a wheel and a sensor.

The same thing could have happened in a car operated by a human driver. And if it had, the human driver might have been inclined to shrug and chalk it up to bad luck. Chauffeur, in contrast, did a whole heck of a lot more than shrug. The team mobilized to make sure nothing like this ever happened again. All told, once they'd completed their tweaks, they'd changed 3,200 parts of the self-driving software, and the new code was uploaded to every one of the team's self-driving cars.

A similar process happened any time the self-driving car acted inappropriately. Which was why the Chauffeur software was getting so good. For example, in March 2017, Uber disclosed that its test drivers disengaged their vehicle's autonomous operation about once a mile. Comparing disengagement rates between companies is tricky, because you never know the difficulty of the roads traveled. Cars traveling on city roads will have a lot higher disengagement rates than those on highways. But just for comparison's sake, that same month, Waymo's disengagement rate was just one per 8,968 miles, or about 8,968 times more reliable than Uber's.

Thanks to the increasingly safe operation of autonomous vehicles, by 2016, with the stampede well under way, I was growing confident that the mobility disruption I'd discussed for so long finally was going to happen—and that we finally were approaching a day that would see the near-elimination of car crashes. And then came the Tesla crash, and for some months in the spring and summer of 2016, the inevitability of the disruption seemed as though it could be slipping away.

Chapter Twelve

HUMAN FACTORS

Subsequent crashes—the March 2018 incident on Route 101 in Mountain View that saw a Tesla Model X in self-driving mode run into a lane divider, killing driver Wei Huang, an Apple engineer, as well as the fatal collision the same month of an Uber autonomous SUV with bicycle-pushing pedestrian Elaine Herzberg in Tempe, Arizona—would provoke similar discussions of safety and reliance on technology. But the Joshua Brown accident was important because it represented the first occasion of its kind.

Tesla's highway driver's assist product was controversial even before the crash happened. The electric vehicle company released the first version of the technology in an update to the Model S software on October 14, 2015. It called the capability "Autopilot." The first Autopilot system was essentially an inferior version of the traffic-jam assist project Chauffeur had aborted back in 2012. It was designed to mitigate the boredom of highway driving by taking over the handling of the vehicle while it navigated well-marked and well-maintained highways accessible only by entrance and exit ramps.

Recall that Google's trials established that, once the autonomous capability becomes engaged, the human driver checks out—and is at high risk of taking too long to return his or her attention to what's happening on the road when a problem presents itself. Some safety experts refer to the issue as the Handoff Problem, with NHTSA's own tests establishing that some drivers require a whopping seventeen seconds to properly take over control of their vehicles, which at highway speeds would translate to a car going about a quarter mile.

Tesla's release of the Autopilot product generated an enormous amount of discussion within the Chauffeur team. We all thought Tesla was being far too aggressive with its marketing of the service. "We've worked at this," Urmson said. "We understand how hard this is. This will not work."

It was only a matter of time, we felt, before someone died in a crash that involved the Autopilot technology. And when that happened, we worried the incident would spawn a backlash—that public sentiment would sour on self-driving vehicles. We also worried a crash could trigger a tighter regulatory atmosphere that would in turn delay the arrival of mobility-on-demand services, exactly the wrong thing to do given the potential of the technology to eliminate the 90 to 95 percent of crashes caused by human error.

So Chauffeur did what it could. The same month that Tesla released Autopilot, the team published an essay on Medium that posed serious questions about highway driver's assist products. The post also explained Chauffeur's reasoning behind the decision to give them up in favor of pursuing full autonomy. "People trust technology very quickly once they see it works. As a result, it's difficult for them to dip in and out of the task of driving when they are encouraged to switch off and relax . . . There's also the challenge of context—once you take back control, do you have enough understanding of what's going on around the vehicle to make the right decision?" The piece ended with what I thought the most effective note: "Everyone thinks getting a car to drive itself is hard. It is. But we suspect it's probably just as hard to get people to pay attention when they're bored or tired and the technology is saying 'don't worry, I've got this . . . for now.'"

To me, Tesla's release was irresponsible. Before Tesla owners were able to use Autopilot, they had to acknowledge that they would maintain hands on the steering wheel and stay alert at all times. They also had to acknowledge that the technology remained in a "beta" stage—that is, it remained an experimental product. Tesla's actions were inherently contradictory and problematic. They suggested to Tesla customers that they check out this cool new product that allows you to not pay attention on the highway—except while you're using it, *you have to pay attention*. And it wasn't just people within the Chauffeur project who thought Tesla's actions were dangerous. Take Zipcar cofounder Robin Chase, who has followed autonomous developments closely since leaving the seminal car-sharing company. "You can't release something like that and then say, 'Yeah, this is imperfect; you have to pay attention,'" she says. "It's too hard."

What many of us expected happened just seven months after Tesla released Autopilot. On May 7, 2016, forty-year-old Joshua Brown became the first fatality in a crash involving the autonomous operation of an automobile—and as we all expected, the incident involved a Tesla Model S using its Autopilot capability.

The technology entrepreneur lived in Canton, Ohio, and ran a business, Nexu Innovations Inc., that specialized in setting up Internet access for clients in difficult-to-serve locations, such as trailer parks.

Before he went into business, Brown was enrolled at the University of New Mexico studying physics and computer science but joined the navy before he obtained his degree. He excelled in the military, serving there for eleven years, becoming a master Explosive Ordnance Disposal technician—a senior-ranked bomb disposal expert. Part of his career was spent in Iraq, disassembling improvised explosive devices and shipping them home to the United States to be examined. He also served for a time with the Naval Special Warfare Development Group, more commonly known by civilians as the Navy's SEAL Team Six.

Another passion of Brown's was his Tesla Model S, which he took ownership of in April 2015, a little more than a year before the fatal crash. He referred to the vehicle as "Tessy," and by May 2016 had already accumulated 45,000 miles on the car. Brown was fascinated enough by Tessy's Autopilot function to post videos of it on his YouTube account. In the seven months between Autopilot's release and the crash, he posted two dozen videos. "Overall it does a fantastically good job," Brown enthuses in one of the videos, in which he explains to viewers the limitations of Autopilot's capabilities navigating hilly and curved back roads—which, for the record, Autopilot was not supposed to navigate.

By far the most viewed of Brown's videos happened in Cleveland as he was driving north in his Model S just as two directions of I-480 merge together to form Jennings Freeway. As chronicled on Tessy's dashcam, Brown was coming up to the Spring Road exit in the right-hand lane when a white truck abruptly swerved across three lanes of traffic. The truck would have swiped the side of Brown's Model S had the Autopilot not reacted. The software first swerved the Model S to the right, to get out of the truck's way, then applied the brakes, allowing the truck to swerve one more time into the exit lane. Brown posted the dashcam footage of the incident to his YouTube account on April 5 under the title "Autopilot Saves Model S."

"[T]he truck," Brown wrote on YouTube, "never saw my Tesla. I actually wasn't watching that direction . . . I became aware of the danger when Tessy alerted me with the 'immediately take over' warning chime"—signaling Brown to resume operation of the car. "It was a mistake on the other driver's part . . . Tessy did great . . . I am very impressed. Excellent job Elon!"

Tesla CEO Elon Musk saw the video and tweeted it out to his millions of followers. Brown was thrilled. "@elonmusk noticed my video," Brown tweeted from his @NexuInnovations account. "With so much testing/driving/talking about it to so many people I'm in 7th heaven!"

The first week of May, Brown vacationed with family members in Orlando, doing things like visiting Disney World. That Saturday, May 7, Brown left to attend a job site in Cedar Key, Florida. After finishing his work there, he set off to drive his Tesla to his next assignment in North Carolina. Investigators believe he took Florida State Road 24 to the town of Bronson, then turned eastbound onto U.S. 27A, which would soon become a four-lane divided highway of the sort that one finds in rural areas, with left-hand turn lanes and intersections that allow vehicles to cross or turn onto the highway after stopping at a stop sign. During the 41 minutes that he was on SR 24 and U.S. 27A, Brown had Autopilot activated for 37 minutes. In that time, Tesla's own software alerted Brown with visual or auditory warnings to resume control of the vehicle on seven different occasions. In response, according to a National Transportation Safety Board report, citing the Tesla's own software, Brown's hands were on the steering wheel for only 25 seconds.

Brown turned on U.S. 27A about 35 minutes into the trip. About 4 minutes later he increased his speed to 74 mph. (U.S. 27A has a speed limit of 65 mph.) One minute and 51 seconds later, at 4:36 P.M. with conditions clear, Brown and his Tesla crested a rise in the road as a tractor-trailer turned left across U.S. 27A. The Tesla didn't stop. It didn't even brake. Brown's car hit the middle of the trailer in the space between the two sets of wheels. The underside of the tractor-trailer tore off Tessy's roof. The airbags didn't deploy because the vehicle speed didn't change. Brown's vehicle passed right under the trailer, veered to the right off the highway and through a series of fields and fences before finally coming to a stop a quarter mile away after it collided with a hydro pole. Brown died instantly in the collision.

Brown's tragic death amounted to the first significant safety-related crisis for the mobility disruption. Few people heard about the accident in its immediate aftermath. The significance of the crash only became apparent more than a month later, in late June, when NHTSA opened an investigation to determine the cause. Had Autopilot sensed the imminent collision? Had it warned Brown? If it had, why hadn't he been able to avoid the collision? And if not, why not?

Many different investigations were conducted into Joshua Brown's death. The Florida Highway Patrol, the National Highway Traffic Safety Administration, the National Transportation Safety Board and numerous media organizations used eyewitness interviews and other research to reconstruct the circumstances of the crash. (Even the Securities and Exchange Commission looked into the matter to discern whether Tesla had informed investors of the crash in a timely manner.) The driver of the truck that pulled in front of Brown, Frank Baressi, told reporters that he didn't see the Tesla.

NHTSA's Special Crash Investigations team conducted a reconstruction of the crash that suggested the tractor-trailer could have been visible to Brown for more than ten seconds before the impact. But the absence of tire skids suggested that Brown wasn't aware of the impending collision, and a later investigation concluded that Brown "took no braking, steering or other actions to avoid the collision." The obvious conclusion is that Brown, for whatever reason, was so confident in the Tesla's self-driving ability that he wasn't paying attention to the road.

But why didn't Tesla's Autopilot notice the trailer? Investigator Kareem Habib of NHTSA's Office of Defects Investigation confirmed that the automatic emergency braking system didn't provide any warning as the vehicle approached the trailer. What good was a technology that couldn't stop when faced with an enormous tractor-trailer that the driver himself could have seen ten seconds before impact?

That summer, Tesla staff is reported to have testified before the U.S. Senate Commerce Committee that the company had two theories. One, that the radar and camera may not have detected the trailer, thinking instead that the bright white surface was simply part of the sky on a bright and sunny day. The other theory was that the Autopilot sensors *had* detected the trailer but incorrectly discounted the structure as a "false positive," an overhead highway traffic sign or a bridge that the vehicle could safely navigate under.

Subsequent investigations by NHTSA and the National Transportation Safety Board revealed that the Tesla's self-driving system simply wasn't designed to protect against situations in which the vehicle approached the sides of other vehicles. Tesla's system was designed for the sort of divided highways that exist in the interstate system, with on-ramps and off-ramps, and medians or walls separating the road from anything traveling perpendicular to the vehicle.

You were supposed to direct the car yourself onto an interstate-class highway and only then activate the system, to eliminate the tedium of highway driving, same as you might cruise control. Such systems work in part by classifying objects around the vehicle. They've been taught to recognize the rear profiles of thousands of different trucks and cars. Notably, at the time of the Brown crash, the Tesla system had not been taught the *side* profiles of trucks or other vehicles. Why would it? On the sort of interstate freeway on which Autopilot was intended to be used, which are designed to avoid intersections altogether, it was hard to consider any scenarios that would confront the technology with the side of a vehicle.

At the time of the crash, Tesla utilized for its self-driving system components designed and manufactured by Israel's Mobileye NV, a supplier of advanced driver assist tools for the automotive industry. And the Mobileye EyeQ3 product used by Tesla did not protect against two main classes of collisions that tend to happen at intersections—what NHTSA calls straight-crossing-path and left-turn-across-path crashes. In each scenario, a vehicle traveling perpendicular to the Autopilot-enabled car blocks the car's path, either by proceeding through an intersection on a road perpendicular to the Autopilot car's road or turning left across its road. In the words of NHTSA's report, " The Florida crash involved a target image (side of a tractor trailer) that would not be a 'true' target in the EyeQ3 vision system dataset."

Consequently, NHTSA's investigator concluded that, fundamentally, Tesla's technology worked fine. "The system's capabilities are in-line with industry state of the art," the investigator wrote. It was just that "braking for crossing path collisions, such as that present in the Florida fatal crash, are outside the expected performance capabilities of the system." This was true of other highway driver's assist products besides Tesla's. None of the similar Mobileye systems in use by other automakers, such as the ones used at the time of the incident by the Infiniti Q50, the Mercedes-Benz S65, the BMW 50i, the Audi A7 or the Volvo XC60, protected against such crashes.

So here's the situation: Tesla rolls out a form of self-driving technology that doesn't protect against two major types of collisions. It takes some steps to protect itself against legal liability. For example, the owner's manual cautions drivers that one component of Autopilot, the Autosteer system, is designed to be used "only on highways and limited-access roads with a fully attentive driver." But how many owners actually encounter that warning? As National Transportation Safety Board member Christopher A. Hart would later observe, the warning "fails to consider the reality that very few owners, and even fewer non-owner drivers, read the manual. Some may look at it only twice a year, to reset the clock when daylight savings time begins and ends." If only Tesla had adequately conveyed the limitations of its vehicle's autonomy.

Tesla set up the circumstances of this crash in numerous different ways, by my estimate. One, and most obviously, it called the suite of technologies Autopilot. In the wake of the crash, Morgan Stanley research analyst Adam Jonas suggested Tesla stop using the name Autopilot, since it "could create a consumer expectation problem and a potential moral hazard." The safety watchdog *Consumer Reports* questioned whether the company "promoted a dangerously premature assumption that the Model S was capable of truly driving on its own.

"Many automakers are introducing this type of semi-autonomous technology into their vehicles at a rapid pace, but Tesla has been uniquely aggressive in its deployment," wrote *Consumer Reports*. "It is the only manufacturer that allows drivers to take their hands off the wheel for significant periods of time, and the fatal crash has brought the potential risks into sharp relief." Then, like Jonas, *Consumer Reports* called on the company to stop referring to the system as Autopilot.

Beyond the name, Tesla didn't adequately inform its users of the system's capabilities. It's reasonable for a driver to assume a vehicle that can adequately handle itself in a traffic jam on an interstate would be able to appropriately react to an obstacle as sizable as a tractor-trailer. At the time of the Brown crash, Tesla suggested that its self-driving products were intended to be used on "limited-access" highways. With Brown thought to have used Autopilot on both SR 24 and U.S. 27, neither of which is a limited-access highway, apparently he didn't understand the limitations.

In retrospect, it seems astonishingly reckless that Tesla would roll out a system that allowed for extended periods of hands-free driving, while that system didn't protect against several major categories of dangerous crashes on roads where the company had to have known its users were employing the technology—thanks to videos like those Brown was posting to YouTube.

— — —

Just weeks after the Brown crash reached public attention, the Israeli company that supplied Tesla with parts for its self-driving systems, Mobileye, severed its relationship with the automaker. The reason, Mobileye chairman and chief technology officer Amnon Shashua explained, was because Tesla was "pushing the envelope in terms of safety."

The Mobileye system that Tesla used in the Model S that crashed "is not designed to cover all possible crash situations in a safe manner," said Shashua to Reuters. "No matter how you spin it, [Autopilot] is not designed for that. It is a driver assistance system and nor a *driverless* system." (My emphasis added.)

Shashua said that in the aftermath of the Joshua Brown crash, Mobileye was reacting against Tesla's mixed messages—mentioning in particular the way Musk and the company bragged about the technology's autonomous capabilities, while warning Tesla owners to pay attention and drive with their hands on the wheel. "Long term, this is going to hurt the interests of the company and hurt the interests of an entire industry, if a company of our reputation will continue to be associated with this type of pushing the envelope in terms of safety," Shashua said.

This was unusually substantive criticism between billion-dollar companies. Tesla responded by insisting that Mobileye was only issuing its criticism because it had learned that Tesla would soon be developing its own computer vision system and was irritated about it. But Mobileye came back hard in response. In a press release, Mobileye recounted that, dating back to May 2015, a year before the Joshua Brown death, Shashua had expressed his "safety concerns" about Autopilot's hands-free use to Musk. "After a subsequent face to face meeting," Mobileye said in the release, "Tesla's CEO confirmed that activation of Autopilot would be 'hands on.'" Despite Musk's assurance to Shashua, the company noted, "Autopilot was rolled out in late 2015 with a hands-free activation mode." Even after that, Mobileye had continued to push Tesla to be more conservative with its Autopilot marketing, the company said.

The same week that Mobileye and Tesla were bickering, Tesla conducted an "over the air" update of its Autopilot software to take extra steps to ensure that its drivers were paying attention to the road. For example, a warning beep would prompt the driver to return hands to the steering wheel after a minute if the vehicle was proceeding faster than 45 mph without any car ahead of it. Ignoring three warnings in an hour would disable the Autopilot until the vehicle was parked. Musk said the update, which also included improved radar capabilities, likely would have prevented the Joshua Brown crash. However, it's doubtful that Mobileye's concerns, or the concerns of other informed observers, would have been assuaged by even these measures.

— — —

I took several things from Joshua Brown's death. Tesla have a history of overpromising and under-delivering. Autopilot was an example of that tendency. As an enthusiastic proponent of self-driving technology, I feel lucky that Larry Page and Sergey Brin funded the development of the technology long before Elon Musk ever automated the driving of his vehicles. Google conducted its self-driving tests on public roads for seven years without its systems being implicated in a single crash—and when it finally did come along, it was a fender bender that occurred at 2 mph. Chauffeur's safety culture and extraordinary team of test drivers and engineers made this possible. In contrast, Musk and Tesla rolled out their test version of Autopilot in October 2015 and just seven *months* later the system factored in a fatality. Can you imagine if the Google team had experienced a fatal crash seven months into its testing back in 2009? The stampede into the market, the excitement around the space—none of that would be happening.

Thankfully, the public outrage over the crash never materialized to the extent that we feared. In fact, each time autonomous technology has factored into a fatal car crash, thanks to sober evaluation by regulators and impressively even-handed coverage by media, the crises have developed into opportunities to educate the public about the current limitations, and exciting future benefits, of self-driving technology.

Epilogue

THE QUEST GOES ON

It ain't over till it's over.

—YOGI BERRA

In October 2017, to coincide with the ten-year anniversary of its win at the DARPA Urban Challenge, Carnegie Mellon University invited the members of its self-driving-car teams back to Pittsburgh for a reunion. The event coincided with a series of panel discussions on the history, present and future of autonomous technology. Red Whittaker was kind enough to invite me to participate. It was a wonderful event, with a special reception held at Pittsburgh's beautiful Phipps Conservatory, and it represented for me, as well as people like Urmson and Salesky and all the others who attended, an opportunity to consider how far we'd all come.

One of the high points was Whittaker's role as master of ceremonies at the reception, where the legendary robotics professor revealed that Spencer Spiker had solved the twelve-year-old mystery behind the engine trouble that caused H1ghlander to come in third in the 2005 DARPA Grand Challenge.

Just days before, Spiker had been crawling over the Humvee's engine to clean it up and make sure it was ready to be displayed on campus during the reunion. His knee accidentally pressed against the electromagnetic interference filter, which reduced the static in the signal to the Humvee's fuel injection system, and the engine died. That was strange, Spiker thought. Subsequent investigation revealed that the filter, which must have been damaged in the rollover, caused the engine's unreliable operation. After the speech, Whittaker handed the filter over to Chris Urmson, who examined the device, which was a little smaller than a Rubik's Cube. "This would have been a useful thing to know twelve years ago," Urmson said. Then he looked up at the man who was his former thesis supervisor and grinned, "I don't think I'd change anything, though. Things worked out pretty well for us all."

That was my takeaway, too. Back in those days, as the races were happening and for years after, the technologists working on self-driving cars, as well as many of the other research projects that had the potential to reduce the waste in the auto industry and transform personal mobility, often exhibited a frustration that society, and the auto industry overall, was either unwilling or unable to understand what was possible. I was as guilty of feeling that frustration as anyone else, when, say, my R&D budget was cut, or the media didn't get the point of a concept car.

I no longer feel that frustration. Nor, I imagine, do any of the engineers and computer scientists who gathered together in Pittsburgh that weekend. Because our predictions have been borne out. We exist in a world now that has changed a lot in just a decade—and will continue to change, just as most of us predicted, and hoped. These days Red Whittaker, who once felt as though he spent a significant amount of time battling with university administration to develop and test his robots, is feted on banners strung up around campus that read, "This one is for the revolutionaries." And Whittaker absolutely remains a revolutionary; about to enter his eighth decade on the planet, he's still pushing the bounds of what's possible, as one of his many start-ups develops a robot to explore the surface of the moon.

More than seven years ago, when I first started working with the Chauffeur project, in 2011, I was struck by the engineering team's dismissal of Detroit's record of innovation. The Silicon Valley roboticists thought the auto industry was tradition-bound, and tired. Similarly, I was frustrated by Detroit's hostility toward Chauffeur's work.

Since then, the attitudes on both sides have changed. Now known as Waymo, the former Chauffeur team has come to respect Detroit—just as the auto industry that once derided their work has embraced a future of autonomous mobility on demand. The enmity that once characterized the relationship between Silicon Valley and Detroit has given way to a spirit of collaboration. I criticized my old company's CEO, Mary Barra, for her proclamation that 2016 would be the year that Detroit "took on Silicon Valley." Barra would soon back down from the combative language. Her team at GM would establish a template for collaboration that would come to dominate the way the major automakers approached the problem of developing driverless mobility by outsourcing the R&D to Silicon Valley technologists. Barra's company created that template with her March 2016 purchase of Cruise Automation. She further strengthened GM's hand by fully engineering the Chevy Bolt EV so that Cruise's self-driving system could be factory-installed equipment and built in high volume when ready. Barra's strategy received a strong endorsement in May 2018 when Japan's SoftBank Group announced a $2.25 billion investment in Cruise, one of the sector's largest-ever deals, with particular significance given SoftBank's reputation as a savvy player in the mobility space. The investment valued Cruise, originally purchased by GM for $581 million, at $11.5 billion—quite an appreciation. Less than a year later, Ford executive John Casesa was instrumental in the creation of that automaker's self-driving start-up, Argo AI, led by Bryan Salesky after he left Google and joined up with his old thesis adviser, Pete Rander, formerly of Uber. Casesa now serves on Argo's board. The Ford board of directors also demonstrated their commitment to the automaker's new direction when it installed as company CEO the head of the Smart Mobility program, Jim Hackett.

Chris Urmson would arrange a similar deal after he left Chauffeur in August 2016 to become CEO of an autonomous vehicle company cofounded with Sterling Anderson, formerly of Tesla, and Drew Bagnall, formerly of Uber. Known as Aurora, and located in Palo Alto, the company would arrange its own collaborative deal with Volkswagen and Hyundai and announce in 2018 that it had secured $90 million in start-up funding. Aurora's values statement reads like Urmson's defiant response to spending seven and a half years working in close proximity to Levandowski: "We do the right thing even if it delays us or makes us less money," it says. "No jerks: We debate and solve hard technical problems. We don't waste time battling over personalities and egos and we have no tolerance for time-wasters and nonsense."

A lot has been made of the way that Chauffeur is said to have paid its engineers to leave once the bonus plan vested. Supposedly, engineers compensated by the bonus plan may have felt they needed to cash out their equity stakes in the Alphabet spin-off in the event that the company's valuation sank below the $4.5 billion at which Google assessed the company at the end of 2015. While around ten members of the original team left—including Dave Ferguson and Jiajun Zhu, whose Nuro, Inc., autonomous delivery start-up recently raised $92 million—I think it even more significant that most of the team stayed, which represented an enthusiastic endorsement of the way the company has thrived under CEO John Krafcik, with technology development spearheaded by Dmitri Dolgov and Mike Montemerlo, and vehicle partnerships driven by Adam Frost. The loyalty also may turn out to be a smart business decision. Morgan Stanley analysts Adam Jonas and Brian Nowak made news when they estimated, more than a year after Google's own $4.5 billion estimate, that if the company is able to capture a percentage of the miles driven around the world, and charge revenues of $1.25 per mile driven, Waymo could be headed toward a valuation of $70 billion.

After Uber bought Otto, Levandowski would replace John Bares as chief of the ride-sharing giant's Advanced Technologies Group. The resulting tumult provided a strong indication that Page, Brin and Thrun had been right all along to resist Levandowski's aggressive approach to AV development, as well as his entreaties to become Chauffeur's CEO. Levandowski's leadership was dominated by mistakes and accidents, such as an autonomous Uber vehicle running a red light on the company's first day of testing in San Francisco. Waymo also discovered the way Levandowski, a month before he left the company, downloaded from its servers about fourteen thousand documents related to self-driving technology. Levandowski subsequently was fired by Uber after declining to co-operate with their legal staff as they fought Waymo's ensuing intel-lectual property lawsuits. (As part of the research for this book, my cowriter contacted Levandowski's personal legal team to request comment about his depiction here. Berkeley attorney Miles Ehrlich passed on the request, but we didn't hear back.) No matter how brilliant you are, no matter how productive you are, if you don't have integrity, and if you don't garner the trust of others, you're going to lose.

Uber CEO Travis Kalanick stepped down from his role shortly after firing Levandowski. The Waymo versus Uber lawsuit was set-tled in February 2018 with Kalanick's successor as Uber CEO, Dara Khosrowshahi, expressing regret over the way Levandowski's hiring played out. "Uber's acquisition of Otto could and should have been handled differently," Khosrowshahi wrote, in a letter released as the case settled, with Uber agreeing to pay 0.34 percent of its equity to Waymo, a stake worth about $238 million at Uber's $70 billion valuation.

Perhaps the worst indictment of Uber's approach at the time happened at about 10:00 P.M. on March 18, 2018, when a forty-nine-year-old woman named Elaine Herzberg pushed a bag-laden bicycle across the four northbound lanes of Mill Avenue in Tempe, Arizona. An Uber-operated Volvo XC90 SUV in autonomous mode, traveling with a safety driver at about 40 mph in a 35-mph zone, collided with Herzberg, killing her. Herzberg became the first pedestrian fatality of a collision with an autonomous vehicle. A sub-sequent NTSB investigation revealed that the Uber SUV's sensors detected Herzberg about 6 seconds before the impact, classifying her unusual bag-laden bicycle profile first as an unknown object, then as a vehicle, and finally as a bicycle. Then, 1.3 seconds before impact, the system decided the situation warranted emergency brak-ing—possibly enough time to prevent Herzberg's death. However, to prevent erratic driving due to false positives, Uber had disabled emergency braking maneuvers and instead relied on its software to alert its human safety driver to take over operation of the SUV. Unfortunately, according to the NTSB, the safety driver didn't react in time, apparently because she was distracted by the vehicle's self-driving interface. She didn't apply the brakes until after the impact. It was yet another dangerous situation created by the handoff problem: that is, the inability of a human operator to take over from the autonomous software. Uber halted its autonomous testing in the crash's immediate aftermath.

It was another historical moment for autonomous vehicles, one that threatened to halt progress toward a wider mobility disruption, just as the Joshua Brown crash had. But then Waymo demonstrated its leadership with a remarkable series of announcements and public appearances; indeed, in the weeks after the Herzberg crash, Waymo was said to have conducted more media relations than it had during the previous nine years it had been known as the Google self-driving-car team.

The week after the Herzberg crash, Krafcik traveled to Las Vegas, where he made headlines when he told the National Automobile Dealers Association that the Waymo technology would have been able to properly handle incidents like the one that killed Elaine Herzberg. Days later, at the New York International Auto Show, Krafcik announced the company's single biggest deal with an automaker—the commitment to incorporate up to 20,000 Jaguar I-PACE electric SUVs as part of the Waymo fleet by the end of 2020. Later deals, such as an agreement with Fiat Chrysler to purchase up to 62,000 Pacifica minivans, would exceed even these numbers, as Waymo executed its expansion strategy.

The announcements amounted to the most comprehensive description yet of the future that Waymo hoped to bring about. Rather than a disruptor of existing businesses, Waymo portrays itself as an enabler of future businesses. One way that's happening? Through partnerships with such companies as Avis and AutoNation, who have committed to maintaining Waymo's growing fleet. According to Krafcik, these partnerships are working well and will help Waymo to scale its business at lower cost, leading to more people enjoying new forms of mobility.

Along with previously disclosed news—the company's millions of miles driven, the fact that Waymo Chrysler Pacificas already were providing autonomous rides in Phoenix without any safety drivers behind the wheel—Krafcik and his company were demonstrating their unwavering commitment to bringing about an autonomous future. "We want to create a driver that never gets drunk, never gets tired, never gets distracted," Krafcik said at one public appearance. Having launched the first driverless commercial transportation service in 2018 in Phoenix, Waymo soon will be providing a million trips per day with a tailored mobility service, meaning lots of different sizes and types of vehicle options. You're taking a carload of kids to soccer practice? Use a Chrysler Pacifica. Going out to dinner? Use a Jaguar I PACE. Taken together, the press appearances served to consolidate the company's position as the clear leader in the self-driving space.

I continue to marvel at the unwavering commitment Larry Page and Sergey Brin have made to developing driverless vehicles. The stakeholders in the 130-year-old roadway transportation system, like auto, oil and insurance companies, would *never* have catalyzed the mobility revolution because they had too much vested in the current system. It took visionaries like Page and Brin, with their belief in the potential of digital technology, their passion for designing compelling experiences, their deep pockets to act on their aspirations and their commitment to make the world a better place, to kick-start the new age of automobility.

This last decade has been a learning process for everyone who works in this space. We've all grown. All changed our minds about one thing or another. How far we've come from those days in Victorville, at the DARPA Urban Challenge, when many of us figured that we could get this thing done—if only a single company would pony up the cash.

Well, lots of companies have ponied up a whole lot of cash, billions of it, in fact, and we're nowhere near done. This thing is harder than it looks. On nearly every front—technological, societal, political—we've all realized the scale of the challenge is much bigger than any of us envisioned. No single company can do it alone.

And yet, the quest goes on. I do think it's inevitable. My favorite symbol that this is happening came the same month as Krafcik's announcement, when the ultimate car guy, Bob Lutz, the helicopter-flying, muscle-car-developing, cigar-chomping and climate-change-denying former vice chairman of General Motors, published an essay in *Automotive News*. "The era of the human-driven automobile, its repair facilities, its dealerships, the media surrounding it—all will be gone in 20 years," Lutz wrote. "The end state will be the fully autonomous module with no capability for the driver to exercise command."

I couldn't believe it—Lutz, age eighty-five, the man at GM who tried countless times to cut the R&D budget I was spending to get us to exactly the future he described, had finally come around.

If that doesn't mean the mobility disruption is inevitable, I don't know what does.

Who is going to win?

Simply to ask that question is to misunderstand the scale of what's happening, and I'm as guilty of that thinking as anyone. A previous iteration of this book's subtitle was "The *Race* to Build the Driverless Car." A race being something with winners and losers. Sure, some people, and companies, will do better than others, and some will fare worse. To me, it's hard to imagine a future that sees the major automakers with their market capitalizations in the tens of billions of dollars ever outright beating the technology companies, like Alphabet, Apple and Amazon, with values that bob in the middle hundred billions. But on some level, this is something that will improve so many things for so many people that I think it behooves us all to help to work toward it.

The mobility disruption will not affect everyone in the same way. As we've mentioned, for the elderly or those who live with disabilities, the prospect of liberated mobility likely will be overwhelmingly positive. Others, though, may lose their jobs as a result of autonomous technologies, or due to the downsized economics of the automotive or oil sector. Many of these people will find work in mobility management, content creation for autonomous-vehicle riders or the manufacturing of fuel cells or electric batteries. Remember, more than a century ago, plenty of people worked as blacksmiths providing horses with shoes—and years later, it turned out most of them navigated the automobile disruption just fine.

There's a lot of speculation about whether children born today will ever get their drivers' licenses. That's a good question. Some might, the way some children today still learn how to ride horseback. But I think by the time a child born today gets old enough to drive, the imperative to drive, for the freedom of it, will have dissolved. Freedom of mobility will exist for everyone, regardless of whether you're able to operate a motor vehicle.

A future without human-driven cars will not be a utopia. Not by itself. Bear in mind how great the science-fiction writers depicted the Internet before anyone considered the medium might also produce Internet trolls, fake news and doxxing. Nevertheless, life will improve after the mobility disruption. When road rage is a thing of the past, and the labor changes have been sorted out, our cities will have rationalized themselves into more pleasant habitats, more appropriate for people, and many of the inconveniences that defined our daily routines will have evaporated.

I'm going to conclude with a joke, which happens to be a trademark of mine.

It starts like this: An old farmer is riding in a wagon towed by his old and nearly blind horse, Buddy. The farmer comes across a stranger whose car is stuck in a rut. The stranger asks the farmer if he would help pull the car out of the rut. The farmer says he will and hitches Buddy to the rear of the car. Then he starts yelling.

"Pull, Ginger, pull!" he shouts, and nothing happens.

"Pull, Coco, pull!" he hollers, and again nothing happens.

"Pull, Daisy, pull!" he yells, and still—nothing.

Finally, the farmer cries as loudly as he can, "Pull, Buddy, pull!" And Buddy pulls the car out of the rut!

The stranger is very grateful—and then he asks the farmer why he called Buddy all those different names?

"Oh," the farmer replies, "Old Buddy can't see nuttin' and if he thought he was pullin' all by himself, he wouldn't even try."

I often felt like Buddy while I tried to pull the auto industry out of its 130-year-old rut. With one crucial difference: Buddy couldn't see too well and I can't hear too well. Many of us felt like we were pulling all by ourselves in those early years of developing these transformative technologies. But now, I think, we all realize how many people out there were pulling for the same thing.

Byron McCormick has been pulling since he graduated from Arizona State University in 1974. Robin Chase has been pulling since she conceived of Zipcar. Martin Eberhard and Marc Tarpenning have been pulling since they teamed up to launch Tesla. Tony Tether pulled hard when he led the DARPA challenges. And Red Whittaker, Sebastian Thrun, Chris Urmson, Bryan Salesky and many more pulled impressively as DARPA Challenge competitors.

Then came people like Elon Musk, Travis Kalanick and, most importantly of all, Larry Page, Sergey Brin and John Krafcik. They all started pulling with much more strength.

Looking back, I can't stop marveling at that moment in Victorville, California, after the DARPA Urban Challenge in 2007—when everything changed. That race set up the battle between incumbents and disruptors that will define the future of the auto industry—and personal mobility in general. At Detroit's darkest hour, you had these bold plays from Google, Tesla, Uber and Lyft. The timing's remarkable.

If we pull it off, and we will, we're going to take 1.3 million fatalities a year and cut them by 90 percent. We're going to eliminate oil dependence in transportation. We're going to erase the challenges of parking in cities. All that land will allow us to reshape downtowns. People who haven't been able to afford a car will be able to afford the sort of mobility only afforded to those with cars. And we're going to slow climate change.

One thing's certain: We're heading for interesting times. Enjoy the ride!

ACKNOWLEDGMENTS

This book exists because of the extraordinary vision, tenacious research and exceptional writing provided by Christopher Shulgan. It was Chris and his brother, Mark, who had the idea to write a book about the quest for driverless cars using me as the narrator. I deeply thank Chris for being an outstanding journalist and a wonderful collaborator.

I also want to thank the key characters in the book who played central roles in creating the future, willingly told their stories and were generous with their time. Included are Chris Urmson, Red Whittaker, Sebastian Thrun, Tony Tether, Bryan Salesky, Rick Wagoner, Bill Jordan, John Casesa and John Krafcik.

I am indebted to my General Motors colleagues Byron McCormick, Christopher Borroni-Bird, Alan Taub and David Vander Veen, who inspired much of what we accomplished while I led GM R&D. In addition to Rick Wagoner, I am also indebted to two other GM mentors, Don Hackworth and Tom Davis, who taught me so much about leadership, manufacturing and engineering, and allowed me to reach my full potential during my GM career.

Tether and his DARPA colleagues and Larry Page and Sergey Brin played major roles in accelerating autonomous vehicle technology. Their vision, leadership and commitment to improving people's lives pulled ahead the realization of the benefits of driverless cars by a decade, if not more. I believe their actions will ultimately save 10 million lives worldwide . . . a tremendous legacy. I also appreciate Larry and Sergey for allowing me to collaborate with Google self-driving cars and Waymo for more than seven years, in what has turned out to be the most exciting and promising technology-development initiative in my career.

I want to thank Jeff Sachs for the opportunity to lead the Program on Sustainable Mobility at Columbia University. He inspired and helped fund my research with Bill Jordan and Bonnie Scarborough that quantified for the first time the enormous economic disruption resulting from the "new age of automobility."

I also appreciate the people who worked to ensure that this book was an accurate and truthful chronicle of the events depicted. Thanks to Robin Chase, John Kyros, Doug Field, Adam Frost, Dmitri Dolgov, Nathaniel Fairfield, Mike Montemerlo, Johnny Luu, Syd Kitson, Daniel Yergin, Bob Lange, Robbie Diamond, Adam Jonas, Scott Corwin, Scott Fosgard, Kevin Peterson, Spencer Spiker, Paul and Susan Urmson, Martial Hebert, Herman Herman, John Dolan, Mickey Struthers and Michele Gittleman for their time being interviewed and their support of this initiative.

Myron and Nancy Shulgan, Mark Shulgan, Johan Willems and Jackson Sattell reviewed drafts of key chapters and enhanced the book in several important ways.

I'm grateful to Denise Oswald, executive editor of Ecco HarperCollins Publishers, for agreeing to publish this book and for her extraordinary editing, patience and unwavering encouragement. Emma Janaskie, Trina Hunn, Sonya Cheuse and Diane Burrowes have been wonderful to work with. The contributions of Chris Bucci, my New York agent at CookeMcDermid literary agency, also are much appreciated.

Finally, to my wife, CeCe, and my daughters, Natalee and Hilary, thanks for your love, your unwavering support of my career and your willingness to let me tell my favorite jokes over and over again. I am blessed to be able to share my life with you.

—Larry Burns

A NOTE ON SOURCES

Living through events is not the same as going back and recounting them after the fact. Telling this story the way it needed to be told required the chronicling of many moments in this book that I did not witness myself. For that reason, my collaborator, Chris Shulgan, and I conducted dozens of interviews with those who were eyewitnesses, with many people throughout this story going to extraordinary effort to ensure that we had all the information we needed.

Another useful event was the Waymo versus Uber trial. As a consultant who has worked with Waymo and the Chauffeur team since 2011, as well as many other clients, I've encountered proprietary aspects of business that I cannot discuss in a consumer book. The bargain that I made with Shulgan is that if an event, an email or a conversation became publicly known, or if the sources involved considered it game for discussion, then we could include it in the book—along with some perspective that my unique position provided. The discovery and actual trial proceedings made public numerous aspects of the story and provided us with a lot more material than we might have otherwise had. In addition, we have drawn from the reporting of the many journalists who have adeptly chronicled various pieces of this story. When our chronicle relied on a previously published source, we have made every effort to note the source within the text. The following are the few instances where additional context might benefit the most interested readers.

INTRODUCTION

Automobile transportation statistics: From the U.S. Energy Information Admin-istration, the U.S. Department of Transportation's National Household Travel Survey and the Environmental Protection Agency.

Car crash statistics: The World Health Organization and the National High-way Traffic Safety Administration.

Larry Page in Ann Arbor: Every chronicle of Page's time at the University of Michigan must start with his wonderful 2009 commencement address. Page's 2014 interview with Charlie Rose also was helpful.

CHAPTER ONE

Urmson and Whittaker in the Atacama: Chris Urmson provided the spine of this story, with additional context provided by David Wettergreen's wonderful blog, *Life in the Atacama*. Additional context provided by interviews with Red Whittaker.

Urmson background: Urmson's parents, Paul and Susan, were helpful in pro-viding context here.

Whittaker background: Lots has been written about Red, but little of it has much biographical information. One valuable source for that was James R. Hagerty's 2011 profile for the *Wall Street Journal*, "A Roboticist's Trip from Mines to the Moon." Going back a ways, also useful was John Markoff's 1991 *New York Times* story, "The Creature That Lives in Pittsburgh," as well as the reporting of Byron Spice while he was science editor at the *Pittsburgh Post-Gazette*.

Welcome to the first meeting of the Red Team: The magazine *Scientific American* em-bedded science reporter W. Wayt Gibbs with Red Team. Gibbs's dedi-cated reporting provided my account with context and color. In particular, Gibbs's March 2004 feature, "A New Race of Robots," provided numerous quotes that I used to illustrate the intense way Whittaker motivated his charges as they prepared for the deadline on December 10, 2003, as well as Red's quotes at his team's first meeting in 2003.

The men and women . . . motley crew: Red Whittaker organized a Red Team reunion at the Pittsburgh restaurant Big Jim's in "The Run," early in the research process, expressly for the benefit of this book, which brought together more than a dozen of the folks who lived the DARPA challenges from 2003 through to 2007. Interviews with numerous members of the team who attended, as well as numerous others who didn't, provided context and color. Red's Race Log on the Red Team's archived website, and the Tartan Racing blog, provided useful context. As well, Kevin Peterson and Spencer Spiker both made themselves available for quick phone conversations and speedy email replies throughout the writing process.

Red's leadership style: Context on this comes from numerous interviews, as well as Wayt Gibbs's reporting and the work of Douglas McGray and his March 2004 article in *Wired* magazine, "The Great Robot Race." I also described moments in the DARPA-challenge chapters that were chronicled in such television shows as NOVA's *The Great Robot Race* and the History Channel's *Million Dollar Challenge.*

Finding Robot City: How Mickey Struthers suggested Red Team move into the LTV Coke Works was recounted by Struthers.

The tenth of December, 2003: Wayt Gibbs's account of this night in *Scientific American* also provided the spine of my portrayal.

Maps had become a crucial component: Conversations with Chris Urmson, Red Whittaker and numerous other team members informed the description of the way Red Team's approach relied on mapping. Also valuable was Red Team's post–DARPA challenge research paper, "High Speed Navigation of Unrehearsed Terrain: Red Team Technology for Grand Challenge 2004."

Nevada Automotive Test Center: Interviews with Spencer Spiker and Kevin Peterson helped color my account of the final days of Sandstorm's testing before the March 2004 race. By late 2017 I'd thought I'd heard everything there was to hear about the event. Kudos to Gimlet Media's wonderful StartUp podcast, *Driverless Cars 1: The Grand Challenge,* for unearthing an anecdote I hadn't yet heard: the story of Red Team in Nevada and the RV's wastewater tank.

The actual race: In addition to the already-cited interviews, *Popular Science* writer Joseph Hooper penned a wonderful account of the race events, "From DARPA Grand Challenge 2004: DARPA's Debacle in the Desert." Also valuable was Alex Davies's "An Oral History of the DARPA Grand Challenge, the Grueling Robot Race That Launched the Self-Driving Car."

CHAPTER TWO

Accident at AM General: The account comes from interviews with Spiker, Peterson and Whittaker.

Thrun background, and the origins of the Stanford DARPA team: Interviews with Sebastian Thrun and Mike Montemerlo formed the main source here. Burkhard Bilger's 2013 *New Yorker* feature, "Auto Correct," provided useful background, as did Mark Harris's September 29, 2017, *Wired* Backchannel blog piece, "God Is a Bot, and Anthony Levandowski Is His Messenger." The *Times'* 2012 magazine feature "Sebastian Thrun—King of the Geeks" also provided useful context, as did the *Pittsburgh Post-Gazette* article that Byron Spice wrote about Thrun's Minerva robot in 1998, "Minerva Goes to Washington."

"I signed up to win the Grand Challenge": This quote from Red Whittaker is from the PBS television show *NOVA*, which aired a documentary on the DARPA challenges on March 28, 2006.

Details on DARPA Grand Challenge 2005: DARPA's March 2006 "Report to Congress" was valuable in providing information about the race and its lead-up.

Team Red at DARPA Grand Challenge 2005: Key details of the mechanics of Red Team's pre-planning process are drawn from the 2005 research paper written by Alexander Gutierrez and others, "Preplanning for High Performance Autonomous Traverse of Desert Terrain Exploiting A priori Knowledge to Optimize Speeds and to Detail Paths."

CHAPTER THREE

Dave Hall and Velodyne background: Alan Ohnsman's 2017 *Forbes* profile, "How a 34-Year-Old Audio Equipment Company Is Leading the Self-Driving Car

Revolution," provided context and biographical details on the fascinating inventor, Dave Hall.

Levandowski setting up Carnegie Mellon's LIDAR: The anecdote about the Velodyne LIDAR demo that saw the device fling itself across the room is from an interview with Chris Urmson.

Levandowski background: My interactions with Levandowski over the years were helpful here, as was the Thrun interview and the Burkhard Bilger *New Yorker* article, "Auto Correct." In addition, our understanding of the life of Anthony Levandowski has been profoundly enriched by the reporting of Mark Harris in such publications as *Wired* and *IEEE Spectrum.* I hope Levandowski feels lucky to have such an indefatigable chronicler of his life as Harris, whose work is compellingly written and enthusiastically researched.

Collision between Boss and test vehicle: The anecdote about Boss rear-ending the test vehicle driven by Bryan Salesky comes from interviews with Salesky and Urmson.

Tartan Racing at the Urban Challenge: We're grateful to John Dolan for sharing with us key excerpts of the journal that he kept to inform his family of his experiences at the historic event.

Details and mechanics of the Urban Challenge: The Director of Defense Research and Engineering report "Prizes for Advanced Technology Achievements: Fiscal Year 2007 Annual Report," January 2008, provided useful information on the Urban Challenge. Key race details were provided by DARPA's own website, which still exists online as a valuable archive, as well as an interview with Tony Tether. Also valuable was the Tartan Racing research paper, "Autonomous Driving in Urban Environments: Boss and the Urban Challenge."

The problem of the Jumbotron: Chronicle here assembled from interviews with Tether, Salesky and Urmson, as well as my own witnessing of the events from the stands.

CHAPTER FOUR

Cost of DARPA Challenges: Information found in DARPA's two reports to Congress, both previously cited: "Prizes for Advanced Technology

Achievements: Fiscal Year 2007 Annual Report" and DARPA's March 2006 "Report to Congress."

Early days of the auto industry: Harold Evans's 2006 book, *They Made America: From the Steam Engine to the Search Engine: Two Centuries of Innovators,* provided key historical details on the inventors and engineers who founded the auto industry at the turn of the twentieth century, as did Daniel Yergin's wonderful 1991 history of the oil industry, *The Prize: The Epic Quest for Oil, Money & Power.* (We went from the updated 2008 edition.) Profoundly helpful for facts and context was David Halberstam's 1986 history of the postwar auto industry, *The Reckoning.*

CHAPTER FIVE

Autonomy, Hy-wire and EN-V: Interviews with Byron McCormick, Chris Borroni-Bird and Rick Wagoner helped to triangulate my own recall of these events, as did old copies of speeches I kept in my files.

CHAPTER SIX

Details leading up to the 2008–2009 recession: Bill Vlasic's 2011 history of the auto industry's greatest-ever crisis, *Once Upon a Car: The Fall and Resurrection of America's Big Three Automakers—GM, Ford, and Chrysler,* provided key details and context.

Segway and PUMA: Triangulating my recollection with an interview with former Segway chief engineer Doug Field was helpful here.

PUMA's terrible reception: In addition to the media sources cited in the text, we're grateful to Scott Fosgard and Chris Borroni-Bird for providing their recollections of this period.

After GM: Useful context provided here by the 2010 book I wrote with Bill Mitchell and Chris Borroni-Bird, *Reinventing the Automobile: Personal Urban Mobility for the 21st Century.*

CHAPTER SEVEN

Caterpillar's post-DARPA self-driving project: Interviews with Bryan Salesky and Chris Urmson.

Ground Truth, and Megan Quinn's chocolate-chip cookies: An interview with Sebastian Thrun was helpful. Quinn mentioned the wonderfully human cookie anecdote in a conversation with Kara Swisher on the Recode podcast *Recode Decode,* the transcript of which was published online on December 25, 2017.

Topcon boxes and 510 Systems: Mark Harris's reporting on Anthony Levandowski's fascinating life story provided this anecdote, principally through the 2017 *Wired* feature, "God Is a Bot, and Anthony Levandowski Is His Messenger." Thrun declined to talk about this period of Levandowski's career, citing ongoing legal action. Court documents filed in the lawsuit between Uber and Waymo helped to provide context on 510 Systems, as did Waymo's Arbitration Demand to Levandowski himself.

Levandowski's self-driving pizza delivery: The book's account of Levandowski's attempt at a driverless pizza delivery on Treasure Island is informed by season one, episode eight of the television show *Prototype This!,* as well as a 2008 feature by Declan McCullagh writing for CNET, "Robotic Prius Takes Itself for a Spin Around SF." Also helpful was Mark Harris's reporting on Levandowski in *Wired* and *IEEE Spectrum,* and Burkhard Bilger's 2013 *New Yorker* feature, "Auto Correct."

Page pushing Thrun to develop self-driving: Source of this story is an interview with Thrun, which saw Sebastian acting out the dialogue.

Self-driving A-Team gathers in Tahoe: We assembled the account of Thrun inviting the biggest minds from the DARPA challenges to his Lake Tahoe chalet through interviews with Chris Urmson, Bryan Salesky, Sebastian Thrun and Mike Montemerlo.

Page and Brin compiling the ten hard drives: Source is an interview with Sebastian Thrun.

Early days of Chauffeur: This material assembled almost wholly from interviews with engineers who eyewitnessed them. For example, Dmitri Dolgov is the source of the scenes in the Shoreline Amphitheatre parking lot, including controlling the Prius by smartphone and the police encounter. Dolgov and Urmson were generous with their time in recounting their attempts to have the self-driving vehicles navigate the Larry1K challenges. Waymo PR chief

Johnny Luu provided a map of the final route, which allowed us to go back and revisit it with Urmson and Dolgov.

Leading up to Markoff's story: Thrun's interview was most helpful for writing this account.

CHAPTER EIGHT

Urmson and Levandowski in Detroit: Source is Urmson.

Thrun meeting Mulally: Source is Thrun interview.

The issue of 510 Systems: Valuable here was an email chain between Thrun and other Chauffeur team members that became public in the Uber/Waymo litigation discovery process. Email is dated March 21, 2011, and is initially between Sebastian Thrun and Dirk Haehnel. Court notation is document 1105-3, filed August 7, 2017.

Three engineers wanted to leave: Source is Chris Urmson.

Chauffeur Bonus Plan: Our account of the plan is cobbled together through court documents and interviews with plan participants, as well as those who did not participate in the plan.

The story of Tesla: Ashlee Vance's wonderful biography of the Tesla CEO, *Elon Musk: Tesla, SpaceX, and the Quest for a Fantastic Future,* informed my account of the electric automaker's early years.

Origins of ride-sharing: Robin Chase's interview on Zipcar and the early days of car-sharing was useful for both facts and context. Background about Lyft's origins is drawn from Ryan Lawler's *TechCrunch* feature, "Lyft-Off: Zimride's Long Road to Overnight Success," as well as Brad Stone's excellent history, *The Upstarts: How Uber, Airbnb, and the Killer Companies of the New Silicon Valley Are Changing the World.* Biz Carson's August 18, 2016, *Business Insider* interview with Kalanick was valuable.

Early days of Uber: Also on the former Uber CEO, I enjoyed Miguel Helft's December 2016 profile, "How Travis Kalanick Is Building the Ultimate Transportation Machine." Kara Swisher's December 2014 *Vanity Fair* profile, "Man and Uber Man," was helpful for the way it plumbed the recesses of Travis Kalanick's mind. We also drew on online videos and transcripts of Swisher's interviews with Sergey Brin and Travis Kalanick.

CHAPTER NINE

Researching "Transforming Personal Mobility": Bill Jordan was generous with his time as we chronicled this research, most of which occurred in the 2010–2011 time period, some of which was updated later. The U.S. Department of Transportation 2009 National Household Travel Survey provided useful data that assisted with our calculations, as did the American Automobile Association's 2012 article "Your Driving Costs: How Much Are You Really Paying to Drive?" and the *NYC Taxicab Fact Book* (2006). Further information about Manhattan taxicabs obtained from the Taxi and Limousine Commission and FareShareNYC data analysis, with additional information provided by Jeff Novich, Charles Komanoff and Aaron Glazer of FareShareNYC. Finally, to calculate the average straight-line distance between random origins and destinations in a given square, we relied on math described in N. Christofides and S. Eilon (1969), "An Algorithm for the Vehicle Dispatching Problem," *Operations Research Quarterly* 20, no. 31, 309–18.

CHAPTER TEN

Chauffeur's dog food: Details on the team's transition away from highway driver's assist provided in interviews with Nathaniel Fairfield, Dmitri Dolgov and Chris Urmson. The twenty-seven-minutes detail is from a keynote presentation Waymo Director of Operations Shaun Stewart gave at Singapore's Innovfest Unbound conference on June 5, 2018.

It was pretty bad: Larry Page's characterization of the relationship between Urmson and Levandowski is quoted from his deposition in the Uber v. Waymo civil case.

Sidecar, Lyft and UberX origins: Stone's *The Upstarts* provided important facts and background for this section.

Salesky returns to Chauffeur: Source on the direct quotes from Urmson in this paragraph is an interview with Bryan Salesky.

Driver labor statistics: Drawn from Bureau of Labor Statistics data, namely, "Occupational Employment and Wages—May 2017."

Automation stats and labor effects of AI: The Boston Consulting Group article is by Hal Sirkin, Michael Zinser and Justin Rose, "The Robotics Revolution: The Next Great Leap in Manufacturing," dated September 23, 2015. The McKinsey Global Institute report is "A Future That Works: Automation, Employment and Productivity," January 2017.

What sort of new jobs?: Lyft cofounder John Zimmer's "room on wheels" quote is from Alex Kantrowitz's January 6, 2017, Buzzfeed article, "Lyft Co-Founder John Zimmer Drives and Dishes on Automation, Car Subscriptions, and Cash."

E-commerce stats: Forrester Research projection on e-commerce growth from Daniel Keyes, "E-Commerce Will Make Up 17% of All US Retail Sales by 2022—and One Company Is the Main Reason," *Business Insider,* August 11, 2017.

Luddite backlash: The Brookings Institution policy brief is Roger Burkhardt and Colin Bradford, "Addressing the Accelerating Labor Market Dislocation from Digitalization," March 2017.

Firefly rollout: The description of John Markoff's ride in Firefly is from his article, "The Google Car Takes a Step Away from Boring," the *New York Times'* Bits blog, May 27, 2014. He recounted his interview with Brin in "Google's Next Phase in Driverless Cars: No Steering Wheel or Brake Pedals," *New York Times,* May 27, 2014. Kara Swisher's take on Firefly is from video posted on the Recode website at "A Joy Ride in Google's New Self-Driving Clown Car (Video)," May 27, 2014. Video from Swisher's Code Conference interview with Brin is online at Recode. Alexis Madrigal's transcript of the Urmson conference call is online at "Google Answers Some of the Pressing Questions About Its Self-Driving Car," *The Atlantic,* May 29, 2014.

Where does Uber fit?: Kalanick's interview with Kara Swisher will go down as one of the most entertaining in the history of the mobility disruption. My account is based on the video archive of the interview on the Recode website, "The $17 Billion Man: Full Code Conference Video of Uber's Travis Kalanick," June 8, 2014. How Kalanick responded to the Firefly rollout is based in part on material drawn from both Travis Kalanick's and David Drummond's depositions in the Uber v. Waymo civil suit.

Uber's October 2014 board meeting: Again, source is both Drummond's and Kalanick's depositions.

Meanwhile, Kalanick monitored Chauffeur: Emails between Kalanick and Drummond, and the details of the March 10, 2015, meeting between Page, Drummond, Kalanick and Michael, all based on documents disclosed in the Uber/Waymo litigation.

John Casesa's Ford hiring: Source is interview with John Casesa.

Uber was making its own preparations: Uber's targeting of NREC is assembled from numerous different sources, some conducted on background. Court documents, including Bares's deposition, contributed. NREC director Herman Herman was generous with his time and providing Shulgan with a tour and the history of the facility. Uber hiring NREC was informed by Clive Thompson's September 11, 2015, feature in the *New York Times Magazine*, "Uber Would Like to Buy Your Robotics Department." To our minds, the best feature on the Uber-NREC story was Mike Ramsey and Douglas MacMillan's *Wall Street Journal* piece, "Carnegie Mellon Reels After Uber Lures Away Researchers," dated May 31, 2015. Also valuable was Josh Lowensohn's account in *The Verge*, "Uber Gutted Carnegie Mellon's Top Robotics Lab to Build Self-Driving Cars," posted online on May 19, 2015.

The biggest about-face: GM's transition to pursuing autonomous vehicles, and the politics behind that, is based on background interviews, as well as *Fast Company*'s Fall 2016 article "Mary Barra Is Remaking GM's Culture—And the Company Itself," and Keith Naughton's October 2015 feature in *Bloomberg Businessweek*, "Can Detroit Beat Google to the Self-Driving Car?"

GM buys Cruise Automation: Although many articles cited a billion-dollar price tag, we went with the $581 million figure from Melissa Burden's

September 29, 2016, *Detroit News* article, "GM's Cruise Automation Team Grows to 100 Employees."

CHAPTER ELEVEN

New and final phase of Levandowski: The Droz quote is from the September 29, 2017, feature by Mark Harris in *Wired* titled "God Is a Bot, and Anthony Levandowski Is His Messenger." Urmson's "we need to fire Anthony" email is presented to Urmson as an exhibit in his deposition in the Uber versus Waymo litigation, document 1565-15 filed September 14, 2017.

Timing of the Ford deal: Source is the May 29, 2017, *Automotive News* story on the Ford–Chauffeur deal by Sharon Silke Carty, "Failed Google Deal Left Fields in the Lurch."

Urmson staged a meeting: Source on the November 17, 2015, meeting that saw Urmson disclose the valuation of Chauffeur to team members included in the bonus plan is the Levandowski deposition, document 1565-13, page 164, filed September 14, 2017, in the Uber v. Waymo lawsuit. However, the actual valuation is redacted in our copy. That the valuation was $4.5 billion is published in the "God Is a Bot, and Anthony Levandowski Is His Messenger" *Wired* Backchannel feature dated September 29, 2017, by Mark Harris. That Google had spent $1.1 billion to develop its self-driving business is from Mark Harris's September 25, 2017, article in *IEEE Spectrum*, "Google Has Spent Over $1.1 Billion on Self-Driving Tech."

Details on the Ford-Google deal: Much of the context provided in Sharon Silke Carty's *Automotive News* story, "Failed Google Deal Left Fields in the Lurch."

If Silicon Valley and Detroit had a baby: The great line about Krafcik is from an *IndustryWeek*'s profile by Laura Putre, "A Reunion in Detroit for Google's Self Driving Guru," dated February 1, 2016. Much of the background on Krafcik derived from our interview with him, including his hiring process at Google. The statistics illustrating Krafcik's role in Hyundai's turnaround are from Danielle Ivory's feature in the *New York Times*, "Chief Will Step Down After Five Years at Hyundai's U.S. Unit," dated December 27, 2013.

Levandowski's friendship with Kalanick: Although we opted to go with a slightly different timing, the source for Levandowski and Kalanick's propensity to walk through San Francisco is Max Chafkin and Mark Bergen's "Fury Road: Did Uber Steal the Driverless Future from Google?," *Bloomberg,* March 16, 2017. Providing other useful context was Mike Isaac's "How Uber and Waymo Ended Up Rivals in the Race for Driverless Cars," *New York Times,* May 17, 2017.

The vesting of Chauffeur Bonus Plan: That the first of the payouts were to be disbursed on December 31, 2015, is from Levandowski's deposition, document 1565-13, filed September 14, 2017, page 165. The amount of Levandowski's first bonus payment is from the same document, page 168. Source that it was among the largest bonuses ever paid out by Google is the Larry Page deposition in the Waymo v. Uber case, document 1068-6, filed August 2, 2017, page 13, pages 23–25. Levandowski's explanations for why he downloaded the fourteen thousand documents is from Kalanick's deposition in the Waymo v. Uber case, document 1088-1, filed August 4, 2017, pages 91–96.

Levandowski's "Chauffeur is broken" email: Source for the January 9, 2016, email that Levandowski sent Larry Page is Recode reporter Johana Bhuiyan's Twitter feed. Bhuiyan was in San Francisco covering the Waymo v. Uber trial and tweeted out an image of it on February 6, 2018, at 11:04 A.M. In addition, Bhuiyan's reporting on Uber and the trial entertained and informed both of us and provided context throughout the book.

Levandowski's musing about trucking to Larry Page: Source is Larry Page's deposition, cited above, page 71.

Bares's email about Levandowski to Uber leadership: Source of Bares's email is Waymo v. Uber document 1219-1, filed August 15, 2017.

Salesky suggests new roles: The email Salesky sent to Levandowski on January 23, 2016, is in the Larry Page deposition, page 195.

Levandowski's "too much BS" email: Levandowski's January 27, 2016, email is mentioned in the Larry Page deposition, page 103.

Bares's notes on Levandowski: Details from Bares's February 4, 2016, conversation with Salesky were widely reported by the journalists covering the Waymo v. Uber trial, with several live-tweeting Bares's testimony. Several mentioned it in stories. For example, Katie Burke's February 6, 2018, story for *Automotive News,* "Former Uber Employee's Notes Provide Window into Internal Tensions."

Background on Robbie Diamond and SAFE: Interview with Robbie Diamond is source.

Chauffeur's first at-fault accident: Google was remarkably transparent in their response to the February 14, 2016, incident, publishing an account of the collision between bus and self-driving car in the "Google Self-Driving Car Project Monthly Report," February 2016.

CHAPTER TWELVE

Chauffeur did what it could: The essay the Chauffeur team posted to Medium appeared in October 2015 under the title "Why We're Aiming for Fully Self-Driving Vehicles."

Details on the life of Joshua Brown: The most helpful article we read was Rachel Abrams and Annalyn Kurtz, "Joshua Brown, Who Died in Self-Driving Accident, Tested Limits of His Tesla," *New York Times,* July 1, 2016. Brown's YouTube channel also was helpful.

The first week of May: Source of Brown's movements in the days and hours leading up to the crash, as well as other key details and context, was "Collision Between a Car Operating with Automated Vehicle Control Systems and a Tractor-Semitrailer Truck Near Williston, Florida May 7, 2016," National Transportation Safety Board, September 12, 2017. The National Highway Traffic Safety Administration's investigation and report by Kareem Habib, closed on January 19, 2017, also provided key details and context.

Questions about Autopilot: Morgan Stanley analyst Adam Jonas's observation about Tesla and moral hazard occurred in a widely quoted research note. The *Consumer Reports* criticism came from "Tesla's Autopilot: Too Much Autonomy Too Soon," *Consumer Reports,* July 14, 2016.

Tesla versus Mobileye: Reuters did a great job reporting the bickering between the two companies. Included in the coverage was Eric Auchard and Tova Cohen, "Mobileye Says Tesla Was 'Pushing the Envelope in Terms of Safety,'" Reuters, September 14, 2016. The following day, Reuters staff is credited with an article about the Tesla response, "Tesla Says Mobileye Balked After Learning Carmaker to Make Own Cameras." Mobileye's response is from the company's own release, "Mobileye Responds to False Allegations," September 16, 2016.

EPILOGUE

Lutz's essay: See Bob Lutz, "Kiss the Good Times Goodbye," *Automotive News,* undated.

INDEX